War without Bodies

War Culture

Edited by
Daniel Leonard Bernardi

Books in this series address the myriad ways in which warfare informs diverse cultural practices, as well as the way cultural practices—from cinema to social media—inform the practice of warfare. They illuminate the insights and limitations of critical theories that describe, explain, and politicize the phenomena of war culture. Traversing both national and intellectual borders, authors from a wide range of fields and disciplines collectively examine the articulation of war, its everyday practices, and its impact on individuals and societies throughout modern history.

Tanine Allison, *Destructive Sublime: World War II in American Film and Media*
Ahmed Al-Rawi, *Cyberwars in the Middle East*
Brenda M. Boyle, *American War Stories*
Brenda M. Boyle and Jeehyun Lim, eds., *Looking Back on the Vietnam War: Twenty-First-Century Perspectives*
Katherine Chandler, *Unmanning: How Humans, Machines and Media Perform Drone Warfare*
Liz Clarke, *The American Girl Goes to War: Women and National Identity in U.S. Silent Film*
Martin A. Danahay, *War without Bodies: Framing Death from the Crimean to the Iraq War*
Jonna Eagle, *Imperial Affects: Sensational Melodrama and the Attractions of American Cinema*
H. Bruce Franklin, *Crash Course: From the Good War to the Forever War*
Aaron Michael Kerner, *Torture Porn in the Wake of 9/11: Horror, Exploitation, and the Cinema of Sensation*
David Kieran and Edwin A. Martini, eds., *At War: The Military and American Culture in the Twentieth Century and Beyond*
Delia Malia Caparoso Konzett, *Hollywood's Hawaii: Race, Nation, and War*
Nan Levinson, *War Is Not a Game: The New Antiwar Soldiers and the Movement They Built*

Matt Sienkiewicz, *The Other Air Force: U.S. Efforts to Reshape Middle Eastern Media Since 9/11*
Jon Simons and John Louis Lucaites, eds., *In/Visible War: The Culture of War in Twenty-First-Century America*
Roger Stahl, *Through the Crosshairs: The Weapon's Eye in Public War Culture*
Mary Douglas Vavrus, *Postfeminist War: Women and the Media-Military-Industrial Complex*
Simon Wendt, ed., *Warring over Valor: How Race and Gender Shaped American Military Heroism in the Twentieth and Twenty-First Centuries*

War without Bodies

Framing Death from the Crimean to the Iraq War

Martin A. Danahay

RUTGERS UNIVERSITY PRESS
NEW BRUNSWICK, CAMDEN, AND NEWARK,
NEW JERSEY, AND LONDON

Library of Congress Cataloging-in-Publication Data
Names: Danahay, Martin A., author.
Title: War without bodies : framing death from the Crimean to the Iraq War / Martin A. Danahay.
Identifiers: LCCN 2021023670 | ISBN 9781978819207 (hardback) | ISBN 9781978819191 (paperback) | ISBN 9781978819214 (epub) | ISBN 9781978819221 (mobi) | ISBN 9781978819238 (pdf)
Subjects: LCSH: War casualties in mass media. | Mass media and war. | War—Moral and ethical aspects.
Classification: LCC P96.W35 D36 2022 | DDC 305.90695—dc23/eng/20211103
LC record available at https://lccn.loc.gov/2021023670

A British Cataloging-in-Publication record for this book is available from the British Library.

∞ The paper used in this publication meets the requirements of the American National Standard for Information Sciences—Permanence of Paper for Printed Library Materials, ANSI Z39.48-1992.

www.rutgersuniversitypress.org

Manufactured in the United States of America

Where we would end a war
another might take as a beginning,
or as an echo of history, recited again.

—Brian Turner, "A Soldier's Arabic"

Contents

War without Bodies

Introduction

TWO PHOTOGRAPHS

From the nineteenth century onward, technology has disseminated images and descriptions of war with increasing volume and speed. Thanks to the invention of the telegraph and photography, nineteenth-century noncombatants had access to firsthand accounts from correspondents and visual images of armies at war. Continuing technological innovations such as television and satellite communication eventually led to images transmitted via live video directly from the battlefield. However, this ready access to more reports and images did not result in the end of war because, despite the proliferation of accounts of violence, their impact was muted by the way in which they were framed. From the British involvement in the Crimean War (1853–1856) to American media coverage of the Gulf (1990–1991) and Iraq War (2003–2011), the framing of war has meant that the deaths of both soldiers and noncombatants were made invisible. These frames included the trope of masculine self-sacrifice as a "good death" in the nineteenth century, the gamification of war from the early twentieth century onward, the focus on psychological rather than bodily damage in the creation of the diagnosis of post-traumatic stress disorder (PTSD), and the derealization of violence at a distance in "drone vision." This framing process helped sustain a war culture that mobilized popular support for military operations and minimized the cost of state-sponsored violence by representing "war without bodies."

Early in the twentieth century, H. G. Wells pronounced that World War I was "the war that will end war," and Ernst Friedrich in *Krieg dem Kriege!* (*War against War!*) (1924) published graphic images of the damage to bodies in the conflict, but wars continued to be fought. Proclamations such as that by Wells and the publication of graphic images by themselves have not been enough to make war unimaginable. War is very much imaginable when the damage to human bodies is made invisible, which is the effect of representing "war without bodies." By this I do not mean that no bodies are involved, but that the absence of bodies or the use of surrogates in representations of war normalizes violence and makes death and destruction unexceptional. In the nineteenth century, words and images represented war as part of masculine identity in the belief that battle was men's biological destiny. More recently, war has become entertainment as it has been either made into a game or rendered acceptable by excluding images of dead bodies in the media. This book traces a history of the ways in which war has been framed to make it thinkable and even natural in the period between two different conflicts, the Crimean War and the U.S.-led wars against Iraq from the Gulf War of 1990–1991, fought initially in Kuwait, to the invasion of Iraq in 2003 that led to a protracted occupation of the country.

These wars are linked by two photographs that echo each other but also show the differences in their historical and ideological contexts. Roger Fenton's *The Valley of the Shadow of Death* (1885) (Figure 1) is an iconic image from the Crimean War, when he used the new technology of photography to document the British army at war for the first time. Sophie Ristelhueber's images in *Fait: Koweit 1991* (1992) and *Eleven Blowups* (2006), especially of a bomb crater in a nondescript valley (Figure 6), seem at first sight to be late twentieth-century versions of Fenton's photograph.[1] However, the contexts for the two images are radically different in that showing an image of a war without bodies in each case was framed by words and other images that dictated how they would be received. Susan Sontag has argued in *Regarding the Pain of Others* (2003) that photographs are decoded according to the existing prejudices of the viewer, so that the image of a dead child can be used to cast one side of a conflict as baby killers or explained away by the other as enemy propaganda (10). Photographs do not speak for themselves, because interpretations of their meaning are determined not only by the prejudices of the viewer but also by the wider universe of texts and images in which they are embedded.

The use of photography in the Crimean War was marked by the transformation of both the conduct and reporting of war. It was the "first industrial war" (Bektas), in which British forces "used mass-produced rifles, exploding shells, sea mines and armoured coastal assault vessels with long-range cannons" (Dooling). It was also the first war in which the telegraph was used for communication; this book focuses on how war was conveyed by media such as the photograph and telegraph, rather than how it was fought. Paul Virilio in *War and Cinema* (1984) has analyzed the way in which the camera was used as an extension of military strategy and "effectively came under the category of weapon" (8). The transmission of images and words is implicated in the representation of war as implicit propaganda and is weaponized in order to mobilize the civilian population. Virilio starts his history of war and cinema in the early twentieth century, but, as he says, he could have "begun in 1854, at the siege of Sebastopol [*sic*] during the Crimean War" (68) but chose World War I instead; I begin my history of visual media and war with Fenton's 1855 photographs during the siege of Sevastopol. While Fenton's imagery was not "weaponized" in the way that Virilio analyzes in World War I, his Crimean photographs reinforced the image of war as a noble, masculine act of self-sacrifice, as I argue in chapter 1.

Virilio claims that "there is no war, then, without representation" (6), a claim that needs clarification because he is not asserting that war would not exist at all without representation, but that war and representation cannot be separated in the period of history that he examines; hence his claim that the camera had the same status as a weapon, or his tracing of the parallel development of aerial photography and military strategy, which was fundamentally altered by this new mode of perception in World War I (*War and Cinema* 11–30). His assertion is similar to that of Judith Butler in *Frames of War* (2009) that there is "no way to separate . . . the material reality of war from those representational regimes through which it operates and which rationalize its own operation" (29), or, in other words, that war needs to be made thinkable and acceptable through a constellation of words and images promoting organized violence that people accept as normal. Such words and images are not necessarily direct propaganda, but rather more subtle forms of representation that take for granted that war is either necessary or unavoidable.

However, Virilio's assertion that there is no war without representation shows the danger of overstating the congruity of communications technology and the military. The telegraph, for instance, is part of the history of what Virilio termed "dromology" in *Speed and Politics* (1977), or the rapid acceleration of war and politics that threatens people's autonomy and their lives; but this technology did not just accelerate warfare—it also ushered in new methods of communication.[2] Paul Armitage points out that "for Virilio military perception in warfare is comparable to civilian perception" ("Paul Virilio"), but this conflates two very different spheres. The telegraph, in addition to its military uses, also enabled the reporting of William Howard Russell and others to reach a wide audience and inform them of the terrible conditions endured by British troops during the Crimean campaign. Like blogs by soldiers during the Iraq War that reported inadequacies in military preparation, such news forced authorities to address problems that hitherto had been seen as one of the unchangeable facets of war, namely attrition off the battlefield in hospitals. Reports from the Crimean War about the conditions in British military hospitals prompted Florence Nightingale to form a corps of nurses and reform hygienic practices (see chapter 2). An overlap between military and civilian uses of technology certainly exists, and the military is embedded in the civilian sphere in both overt and more subtle ways, but the relationship is more nuanced than Virilio's terms allow.

Virilio frames technology as a determining rather than an enabling medium. Douglas Kellner sees this as problematic, arguing that Virilio has "a flawed conception of technology that is excessively negative and one-sided" ("Virilio, War and Technology"). The same problem can be found in the pioneering work of

Marshall McLuhan and the cultural criticism of Friedrich Kittler. McLuhan, in *Understanding Media* (2004), argued that technology rather than the content dictates the "message" received by an audience and used a light bulb as an example (57). However, because a light bulb it is not trying to persuade an audience, and it can have more than one message, it is a problematic example. McLuhan was not to know this of course, but the incandescent light bulb he used as an example would be succeeded by LED technology, and this would become the center of a political controversy when it was linked to "light bulb socialism" (Haraldsson). Far from having one message, the light bulb could have very different meanings according to political beliefs. McLuhan also does not differentiate between technology and media, in a semantic slippage that allows him to ignore ideological content. Media have more than one message and are often ideologically contradictory; a classic example was the bumper sticker "freedom isn't free," which, along with "support our troops," was a common sight in the U.S. during the Iraq War (Lilley et al. 313–314). The message seems contradictory unless you understand that the "not free" part does not refer to liberty but is meant as a justification for military operations and the "cost" in lives of the conflict. This bumper sticker aligned civilians with the military's Operation Iraqi Freedom and showed political support for the argument that the invasion was protecting American "freedoms" and liberating Iraqis from Saddam Hussein.[3]

If you substitute the word "effect" for "message" in McLuhan's writing, the meaning is not changed, which shows that he is discussing social change rather than communication. For instance, he says that "the 'message' is the change of scale or pace that it introduces into human affairs" (8); McLuhan is arguing here that technology is equivalent to social change. His discussion of the light bulb follows the same pattern when he says "the message of the light bulb is total change" (57); in this sentence you could substitute "the effect of the light bulb" and not change the meaning. It is certainly true that new technology brings about some changes in how people interact, but to call it a "message" implies communication, when he is actually discussing social practices that are never specified. In this book I am interested in how new communications technologies influence how war is represented; unlike McLuhan, I do not believe there is any inherent message in a particular technology, but rather that it is adapted to prevailing norms and may or may not disrupt established ideologies. Even when texts and images are marshaled to protest the effects of war, they may be subverted by a wider visual regime that legitimizes violence, as is the case with Ristelhueber's photographs of the post-battle landscape in Kuwait, for instance (see chapter 5).

A further confusion arises between technology and consciousness in McLuhan's writing. He uses the word "extension" in his subtitle, which suggests a prosthesis, but he is actually arguing that human consciousness is changed by technology (although he offers no evidence for this assertion). Electricity "may be said to have outered the central nervous system itself, including the brain" (269), he argues, as if consciousness itself were embedded in electrically powered technology. He also uses the word "organic" (269) to describe electricity, as if it were itself a body, which in turn allows him to use a word like "extension" for both technology and human bodies. Both the human body and technology are "organic" in his argument, which allows him to fuse them and argue that technology "extends" human senses in a rhetorical sleight of hand. McLuhan assumes a free-floating consciousness with no reference to bodies or material practices.

Both McLuhan and Kittler posit that new media bring about effects at the level of the subject. Kittler's analysis parallels Virilio's in his assertion that the "transport of pictures" in cinema repeats the "transport of bullets" in war (*Gramophone* 124). Kittler, in *Gramophone, Film, Typewriter*, ascribes extraordinary power to such new media to alter how people think, arguing that their effect is to remake consciousness as a machine in "writing and reading, storing and scanning, recording and replaying" (33). Kittler argues that technology alters consciousness fundamentally by making the brain an "information machine" (189), in a piece of hyperbole that both overstates the case and eradicates the affective dimensions of media imagery in its emphasis on the mechanical and material. New media undoubtedly introduce new ways of imagining consciousness, so that with the invention of the computer, for example, words like "input" and "data" came to be widely used to describe how people received and transmitted information, but this new vocabulary signaled a change at the level of representation of thinking rather than thought processes themselves. Kittler updates Descartes's argument *Discourse on Method* that animals were automata and applies the same logic to humans based on an observation of "moving machines fabricated by human industry" (Part 5). The new vocabulary did not signal the dawn of a new computer consciousness, but instead acted as an index of the rising power of technology companies as their products became more ubiquitous, while older technologies like the landline telephone lost their dominance (although terms from the previous technology persist in language such as "wireless," which references older technologies that needed material connections). Kittler makes both technology and the military dominant and presents a teleological argument that cybernetics is the culmination of an inevitable evolutionary process that leads to "a world of the machine" (262). Kittler presents a pessimistic vision much like that in the *Terminator* films, of humans dominated by military technology. Kittler's analysis, especially his argument that the Enigma and COLOSSUS machines fought the real battle of the Second World War, is better suited to wars of the future than of the past, as I argue in the conclusion.

Geoffrey Winthrop-Young has argued that war is the determining factor in Kittler's methodology, but admits that this "depends on a terminological vagueness" (838) about what "war" actually means (839). This is a crucial insight both for Winthrop-Young's analysis and for this book, because the meaning of "war" has changed radically from the Crimean to the Gulf and Iraq wars. For Victorians, "war" was an organized conflict between male soldiers, whereas now it seems as much a state of mind as violence between adversaries. Civil society has become so suffused with the technology and the rhetoric of war that the term has come to "mean so many different things" (Winthrop-Young 839), and is applied to so many different contexts, that it has become unmoored from any reference to state violence. According to Winthrop-Young, "war is now the normal state of affairs for the industrialized nations," which means that "potentially all media technologies, no matter who designed them for what purpose, add to the suffusion of society with war" (837). This is the effect of the "mediatization" of war, "by which warfare is increasingly embedded in and penetrated by media" (Hoskins and O'Loughlin 1323), so that news coverage, for instance, becomes an extension of a military campaign that pervades everyday life. Winthrop-Young's analysis parallels the more focused discussion of U.S. civil society by Jon Simons and John Louis Lucaites in *In/Visible War* (2017) that the military has become both ubiquitous and unremarkable (3). The military use of technology in American society, for instance, is a reflection of the enormous spending power of the military, which has increased almost every year from 1960 to 2020 ("U.S. Military Spending").

How new technologies are deployed depends on existing power relations. In the Victorian era, the telegraph enabled faster communications during military campaigns, but it also helped an ascendant middle class challenge the hegemony of the upper classes in leadership roles in the army thanks to reporting on aristocratic incompetence in the Crimean War (Dawson 107; Figes 469). More broadly, the telegraph enabled a cadre of white, male, middle-class men to become entrepreneurs and challenge upper-class control of capital (Müller 8–9, 46–47). The new technologies, however, did not fundamentally alter the way that war was "framed." Similarly, live video feeds from the front lines allowed 24/7 access to images of war, but such imagery was anesthetized by the voluntary censorship of news organizations in not showing any dead bodies, only material damage.

Framing Death

Words and photographs are embedded in discourses circulating in the wider culture, and frame analysis decodes their context. Framing theory is used in literary analysis, media studies, and political science to discuss the way in which events are presented to the public, especially in terms of bias or ideology; framing theory is also increasingly being used to analyze war reporting (King and Lester 626). In framing, "strands of information are selected, emphasized, linked, and then molded into a coherent tale" (King and Wells 7). Erving Goffman's *Frame Analysis* (1974) is a primary text for this heuristic, and Robert M. Entman's definition of "framing" is frequently cited in the field: "*to select some aspects of a perceived reality and make them more salient in a communicating text, in such a way as to promote a particular problem definition, causal interpretation, moral evaluation and/or treatment recommendation for the item described*" (Entman 52; italics in original). Taking Entman's definition as a baseline and modifying it for my purposes, I define "framing" as the way in which texts and images promote a particular view of war that excludes certain intellectual and emotional responses. While it may not have been the intention of the producer of the text or image to exclude such responses, the cumulative effect of the "framing" of war, especially in media coverage, delimits how texts and images are perceived.[4] Roger Stahl calls this process "framing devices that construct the ritual of watching" (*Through the Crosshairs* 26), which encourages identification with the weapon rather than the "targets" in coverage of drone strikes, for instance.

The concept of "frames" is central to Judith Butler's *Frames of War* (2009), in which she argues that certain lives are made "ungrievable" by being excluded from representation and thus unacknowledged.[5] Butler lists the various ways in which "frame" may be understood and comments that it "implicitly guides the interpretation" (8) of an image, underlining a deliberate process of inclusion and exclusion. This is a "tacit interpretative scheme" (51) that limits the field of representation. Isaac Speer has analyzed how different frames were used to control media coverage of the Iraq War in explicit and indirect ways to ensure support for the war (282–283). The framing occurs in popular media as well as news sources. Drawing on this concept, Holger Pötzsch analyzes the film *Black Hawk Down* (2001) to demonstrate the role of this discursive formation in "reducing the paradigm for possible articulations of both producers and receivers of mass mediated messages" ("Borders" 89). Framing delimits the range of possible reactions to the representation of war in similar ways in fiction and nonfiction.

The same process operates in a more subtle sense when the choice of words guides interpretation. Thus, Alfred, Lord Tennyson, in writing his poem commemorating the charge of the Light Brigade, focused solely on the British cavalry and excluded foreigners such as the Turkish forces who bore the initial brunt of the Russian attack, or the French cavalry who came to the rescue of the retreating remains of the brigade (see chapter 1). Such a focus seemed natural to Tennyson and his readers because of the nationalist assumption that only British losses were "grievable." The deaths of the British cavalry in turn were made acceptable by the wider frame of male self-sacrifice in war as a fulfilment of their masculine identity.

Butler's analysis corroborates that of Sontag, who argued that "to photograph is to frame, and to frame is to exclude" (*Regarding* 41). Butler pushes this idea further by arguing that the way images and texts are framed interpellates a subjectivity that accepts violence as the price of security.[6] Framing thus reinforces ideologies in which the lives of others are deemed of less value than, for example, those of American soldiers versus Iraqi civilians. This does not mean that Americans are necessarily overtly hostile to Iraqis (although some may well be) but that they accept the idea that their security depends on waging war across the globe via preemptive strikes against people labeled as threats, especially if the targets are accused of being terrorists, and that the civilians who die in the attacks are not acknowledged as important and thus are rendered invisible. Iraqi civilians were excluded by the narrative frame and "de-humanized and de-subjectified" (Pötzsch, "Borders" 78).

Butler indexes the different values placed on lives in terms of "precarity" and "precariousness" (*Frames of War* 25) in that some populations are made deliberately vulnerable and subject to violence (26), either by domestic policies or a foreign policy that uses war as a mechanism of power. Butler's analysis helps underscore how "frames structure modes of recognition, especially during times of war" (24) and thus mobilize popular support for violence. In a strangely abstract and distant formulation, Butler asserts that there is "no way to separate . . . the material reality of war from those representational regimes through which it operates and which rationalize its own operation" (29). While this is undoubtedly true, it begs the question of how exactly representations rationalize war, which is the subject of this book. The effect of many representations is precisely to try to separate the reality of war from its imagery, and through this derealization make depictions of violence palatable. Virilio in *War and Cinema* has analyzed "a growing derealization of military conflict" (1) that has accompanied the invention of visual technology and has increased the circulation of images of war, but also made them insubstantial.[7] Similarly, Lilie Chouliaraki's analysis of the mediating role of television images of suffering argues that they "render the spectacle of suffering not only comprehensible but also ethically acceptable for the spectator" (3). Making war an acceptable spectacle makes violence palatable and just another piece of information learned in "a space of safety" (4) for the viewer.[8] Chouliaraki emphasizes the "hierarchical zones of viewing" (4), in which viewers of images of violence are insulated from the emotional effect by the way in which they are framed as happening in a zone of precarity far removed from the safe space that they occupy.

Butler's abstract analysis is made corporeal by Julia Welland, who argues that "there has been a move away from grand and abstract theorizing about war, towards an understanding of war that places the human body and embodied experiences at the centre of theorizing" (528). As she says, "bodies have often remained absent from analytic engagements with war," despite their centrality to promulgating violence (529).[9] Welland here is building on Elaine Scarry's definition in *The Body in Pain* of war as injuring (63) and her argument that the pain and suffering are disavowed because its subjective experience cannot be adequately conveyed in language (60). In discussing the differences and similarities between accounts of war and torture as damaging to bodies, Scarry asserts that both swerve from acknowledging their violence (64); her theory explains how a representation of a "war without bodies" evacuates images of suffering from the frame and makes war acceptable. Theoretically, a focus on damage to the body in war would correct for such avoidance, but whose body and how it is represented are crucial for the effect, or lack of it, on those reading or viewing accounts of violence. If the body is defined as an "enemy," then damage and death are not grievable, which subverts the empathetic response to the suffering of another person. In the occupation of Iraq, for instance, media attention was focused on the psychological well-being of returning American soldiers, and the subsequent omission of civilian body counts made Iraqi deaths ungrievable (see chapter 4). This derealization of the body was accomplished during the Crimean War through a series of surrogates for the soldier's body, such as the horse and the saber (see chapter 1).

As Sontag illustrates through her examination of photographs, there is no intrinsic emotional reaction to an image; context is crucial. The way that images and texts are framed elicits a response "by virtue of the structuring constraints of genre and form on the communicability of affect" (67). Barbie Zelizer, in "When War Is Reduced to a Photograph" (2004), argues that "only certain aspects of war are ever seen in the images of war. Lacking depictions are those sides of war which do not fit the prevailing interpretive assumptions about how war is to be waged" (116). Assumptions about warfare as masculine (see chapter 4) exclude women from coverage, so that Christina Lamb in *Our Bodies, Their Battlefields* (2020) describes how her fellow reporters did not quote a single Iraqi woman, so that "it was as if they weren't there" (4). Even more disturbingly, she documents how rape was tacitly accepted as part of war, so that "it has been the world's most neglected war crime" (7). The framing of war as masculine and the silence about rape prevented the stories that she and others wrote from being printed.

I would add to this exclusion of certain narratives that a cumulative effect is in play here, because previous representations create expectations about how issues will be framed in the future. Thus, Fenton's photograph of the *Valley of the Shadow of Death* was interpreted within an existing set of assumptions about masculine heroism and self-sacrifice and Ristelhueber's aerial photographs of damage from war in the Kuwaiti desert (discussed in chapter 4) circulated alongside images of drone strikes that had the opposite emotional effect of her images. While grief might be an expected emotional response to Fenton's photograph and outrage at the environmental damage to Ristelhueber's photographs of a scarred landscape, such reactions may be subverted by expectations and exposure to other related texts, or in other words by the constraints of genre.

The discussion of bodies and pain needs to be located more specifically in historical examples. Whether damage to the body is represented or not depends on the medium used and what is deemed acceptable in that period, "acceptable" often meaning not inimical to the representational status quo in which war is promulgated but its consequences repressed. During the Crimean War, for instance, the *Illustrated London News* showed no compunction in publishing drawings of dead soldiers, while they were absent from Fenton's photographs because he followed the conventions of portraiture and chose live subjects and therefore did not challenge conventional representations of the military as heroic and manly, partly because of the constraints of genre. In the late twentieth and early twenty-first centuries, bodies were absent from American media coverage of the conflict in Iraq, but in virtual, gamified wars, gore and death were featured plentifully. This is contradictory but not unusual; Welland, for instance, analyzes the representation of bodies in the Imperial War Museum *War Story* exhibition, in which, alongside the "hypervisibility" of British soldiers' bodies (524), civilian casualties were excluded (534). How the image of the damaged body aligns with existing ideologies in each historical context determines who is represented and who is excluded from representation in installations like *War Story*.

Scarry cites Karl Von Clausewitz in *On War* (1832) as the most brutally honest text about warfare, because of its recognition that war's primary purpose is inflicting injury and death on enemy soldiers (Scarry 65). Clausewitz is indeed explicit about the aims of war, but he also recognizes techniques that persuade citizens to support war that he labels "political." Clausewitz does not specify what he means by "political," but his text shows an awareness that, while a society may have the means to go to war, an ideological climate must be created that makes war imaginable and acceptable. Clausewitz glosses "political" as "the standard for determining both the aim of the military force and also the amount of effort to be made" (33), making it the engine that drives the machinery of war. The political for him is crucial in that "the war of nations always starts from a political condition and is called forth by a political motive. It is therefore a political act" (39–40). Daniel Pick in *War Machine* (1993) glosses Clausewitz as arguing that "the sword itself is subordinate to the pen," because "war is an instrument of policy" (32). Clausewitz is discussing the creation of a climate that supports war, although he says that, because of the complexity and uniqueness of each nation's culture, he is unable to make generalizations about how this is achieved in light of its "chameleon-like character" (42). He does, however, say that "we can, therefore, only admit the political as the measure, by considering it in its effect on the masses which it is to move" (33), so that the political is crucial before the outbreak of hostilities because it acts on the civilian population to goad them into supporting war; therefore, argues Clausewitz, "the governance of populations is central to the war effort" (Sylvester 177), and lack of popular support undermines the case for hostilities. Pro-war propaganda, both overt and implicit, is therefore a prelude to violence and key to ensuring popular support for the war. This chain of reasoning leads to Clausewitz's most famous formulation, that "war is a mere continuation of policy by other means" (40), a statement that fuses domestic policy, foreign relations, and the military through governmentality because the population is made part of the wider war effort by its alignment with the state's objectives.[10]

Clausewitz's analysis was prescient and helps me understand a "movement of the masses" toward war that I experienced. I was living in Texas when the Iraq War started in 2003. Previously the United States had been part of an international coalition that invaded Afghanistan to attack the Taliban, the group that had harbored Osama bin Laden and the organizers of the September 11, 2001, attack on the World Trade Center in New York City and the Pentagon in Washington D.C., a combined military action that had widespread international and domestic support. The second invasion of Iraq took place under very different circumstances than that of Afghanistan; at the international level, far fewer countries joined forces with the United States, and some opposed the invasion altogether, most notably France, which became the object of ire and the subject of the derisive label "cheese-eating surrender monkeys" (Younge and Henley). The rationale for invading Iraq was not clear, but the U.S. administration made claims about an imminent threat to the country by Saddam Hussein and his forces as a way to "move the masses," to use Clausewitz's terminology; this included the assertion by then Vice President Dick Cheney that Iraq had "weapons of mass destruction" as one of several rationales for the conflict that were dubious at best (Kessler). Most dramatic was the fear voiced by then Secretary of State Condoleezza Rice that there could be mushroom clouds over American cities after terrorists detonated an atomic weapon (although how they would have acquired such weapons was never explained) (Blitzer). These assertions were part of a wider "visual focus that calls the viewer into place as a national citizen who is part of a unified and patriotic national community" (Apel 151). The strategy framed public discourse so that anyone questioning the rationale for war, either at the local or national level, was branded as unpatriotic (Borger).

On the ground in Texas, the effect of this propaganda effort in favor of war was reflected in indirect ways. A group in my neighborhood went door to door giving out free American flags to demonstrate patriotic support for the war, and "support our troops" bumper stickers became ubiquitous, as did anti-French ones to express displeasure at the French for refusing to join the coalition of forces attacking Iraq; at its most absurd, there was an effort to rename French fries "freedom fries" (BBC News). The constellation of words and images in the media created what Pötzsch terms "an audio-visual war culture" ("Borders" 89) that functioned as "the diegetic justification of war and violence" (78).[11] My act of resistance was to buy a United Nations flag for my car, but it was far outnumbered by the American flags attached to cars in my neighborhood.

The overall result was a pervasive "war mood" (Adelman 153) that fostered unquestioning support for American foreign policy at an emotional level that was impervious to arguments that there was in fact no basis for attacking Iraq as a response to the 9/11 attacks. Erica King and Robert A. Wells note in *Framing the Iraq War Endgame* (2009) how any counter-narrative to the push for the invasion of Iraq was overwhelmed by the Bush administration's bellicose rhetoric and the parameters set on debate by the media (43–44). Adelman has described the way in which emotions "operate in tandem with ideological formations" (6) so that civilian populations are enlisted to support the war at a visceral level. Julian Reid situates this process within the Foucauldian heuristic of biopolitics and the way in which "entire populations are mobilized for the purpose of wholesale slaughter in the name of life necessity" (28); the "necessity" in this case was the fictional threat of nuclear explosions over American cities.[12] Scarry terms war "fiction generating" (140), by which she means that to justify war, a compelling narrative must be generated, but in the context of the invasion of Iraq this takes on an added dimension in that the rationales for war were in retrospect proved to be fictional. The repeated admonition to "support our troops" also subverted resistance to warfare. Adelman has analyzed how the various expressions of gratitude to the troops during this period served to "constrain in advance the . . . options for reply" (154). These expressions of gratitude also served to erase the individuality of the troops and insulate civilians from any need to do more than pay generalized lip service to the ideals of sacrifice for the nation that normalized death and destruction in Iraq.

War Culture

The term "war" itself has become embedded in American discourse, thanks initially to the "War on Poverty" and later the "War on Drugs." Rather than use the term "campaign," the recourse to the vocabulary of war militarizes social policy and culture (McSorley 2) and functions as an unacknowledged metaphor for engagement with the civilian population. Formulations such as the "war on drugs" excise the human presence from the equation, as if drugs themselves were independent agents and not substances used by people.[13] Here language acts like the media coverage of the various Gulf conflicts that did not show dead bodies by erasing them as a mediating term.[14] Scarry refers to the "referential instability" (124) of war, by which she means it can be attached to any number of ideologies such as nationalism and religion, but her suggestive formulation also indicates how the word can be yoked to domestic agendas such as drug policy or campaigns against poverty. Clausewitz's declaration that "war is a mere continuation of policy by other means" (40) could in this context be glossed as war being a continuation of American domestic policy by other means. The "referential instability" of war as a signifier in U.S. political discourse allows the state to mobilize violence domestically in much the same way as it does in foreign policy; the militarization of police forces in the United States is another index of this fusing of international military intervention and domestic policies.[15] The ubiquity, yet invisibility, of this rhetoric is the overall argument of Simons and Lucaites in *In/Visible War* that "war is both hyperreal and unnoticed" (3) in American culture because it has become such an integral part of everyday discourse. Appeals to war as social policy help make violence part of the verbal landscape, just as the "war of images" after the attack on the World Trade Center metastasized through the body politic.[16] Using war as a metaphor for domestic social programs also embeds militaristic methods into internal policies, and is reflected in such developments as the increased militarization of the police (Simons and Lucaites 2–3, 222).

The word "war" was applied to the Iraq conflict, but organized hostilities between American and Iraqi forces lasted a very short time under the "shock and awe" campaign unleashed on March 22, 2003 (CNN), after which it became "asymmetrical warfare."[17] The initial code name for the invasion was Operation Iraqi Freedom, so that it was termed an "operation" and not a war, and the conceit was that Iraqis would greet American forces as liberators, as then Vice President Cheney claimed (Russert). However, the "war" rapidly became an occupation, although it was never named as such, and put American forces in an untenable position, where they were supposed to protect Iraqi civilians but frequently could not tell combatants and noncombatants apart. This created a situation in which the soldiers experienced guilt and remorse because they may have inadvertently killed an innocent person (see chapter 4). Eventually Operation Iraqi Freedom was renamed Operation New Dawn (Jaffe) as the U.S. forces prepared to leave and hand control of the country back to the Iraqi government and its armed forces, but this did not immediately solve the dilemma on the ground for American troops.

Adriana Cavarero has analyzed the way in which the word war itself has become "equivocal and slippery" and "linguistically chaotic" (2), but most disturbingly argues that it masks the "mass homicide of civilians" (62). While the word still conjures up images of soldiers in uniform doing battle with each other in a defined geographical area, which is what it referred to in the Crimean War, the promulgation of warfare has become a constant series of attacks by drones or missiles on selected targets, with civilian deaths underreported or completely ignored; indeed, one estimate puts the casualty rate of civilians due to armed conflict at 90 percent of the total of casualties, with 75 percent estimated to be women and children (Fuller 59). As Stahl says, in media coverage of the Gulf and Iraq wars, "target bodies withered away alongside soldier bodies," so that no bodies were visible outside the "bomb's eye view" that excluded them from the frame (*Through the Crosshairs* 31). This indeterminacy in the language of warfare is exacerbated by the use of drones, which destabilize previous understandings of the boundaries of the conflict, so that "we can call this situation 'war,' but it is no longer clear exactly what that means" (Kahn 199). Cavarero's solution to the exclusion of the real casualties is to rename war "horror" and "horrorism" (3), because it is "violence against the helpless" (29) rather than the traditional understanding of battles between uniformed armies.

When war is framed as both inevitable and unavoidable, as Butler points out, the resulting anxiety generates "specific ontologies of the subject" (*Frames* 3). In Texas, because of this framing, people saw support for the "war on terror" as giving them security in exchange for the loss of certain freedoms to counter "the hovering possibility of a violent happening" (Adelman and Kozol 103). The implicit promise was that they would be secure if they acquiesced to new methods of policing and surveillance. The myth of complete security denies what both Cavarero (31) and Butler see as a fundamental precarity in that human bodies are always at risk from incursions (29–30) and enables the creation of a war culture. The emphasis on terror reinforces the sense of vulnerability for the civilian population, especially if mushroom clouds over U.S. cities are evoked as a possibility. The obverse of this is the projection of an invulnerable military defending civilians thanks to advanced technology that replaces the human presence and obviates civilian casualties. The representation of a "murderously destructive yet simultaneously corpseless" (Norris 230) narrative enables the fiction that the violence is safely quarantined "over there" and that the American public is not complicit in the killing of innocent victims on their behalf. The images created by various artists cited in chapter 5 work to undermine this complacent attitude and to reframe war in a way that shows civilian casualties and the vulnerability of the human body.

The rhetoric and imagery of war changed from the Crimean War to the Gulf War, but it has in common the framing of war in ways that make the loss of life, whether of soldiers or civilians, ungrievable. The chapters in this book thus trace a genealogy of the different ways in which war has been framed to create the illusion of a "war without bodies." This is the process of "substitution" described by Virilio,[18] but where he uses the examples of cinema and digital media, I cite metonyms such as the word "saber" or toy soldiers as a surrogates for the soldier's body, or a focus on the mental instead of physical wounds of battle that diverts attention from civilian casualties.

Outline of Chapters

Chapter 1 examines Fenton's photograph in the context of the Crimean War (1853–1856) and the ideology of the sacrificial male warrior that was reinforced by Victorian concepts of "duty" as obedience. The next chapter examines the shift in the representation of soldiers' bodies after the Crimean War in the wake of newspaper reports describing the awful conditions endured by the troops in dispatches by William Howard Russell and the mythos that grew around Florence Nightingale as "the lady with the lamp." The representation of soldiers shifted to images of the wounded body, as shown by the Crimean War monument in Waterloo Place, London (1861), and paintings by Elizabeth Thompson, Lady Butler, such as *The Roll Call* (1874) and *Balaclava* (1876). The image of the wounded soldier was used as a vehicle to come to terms with the inconclusive end of the Crimean War and implicitly to justify the sacrifices made in a campaign that did not produce a conclusive victory.

The third chapter bridges the Crimean and Gulf and Iraq wars by tracing the history of the gamification of war. While many studies have been published on the militainment of video games, none have traced the genealogy of such gamification. Beginning with H. G. Wells's *Floor Games* (1911) and *Little Wars* (1913), the chapter links the texts to the production of toy soldiers, especially by Britains Ltd., and the development of *Kriegsspiel* from a military strategy game into a civilian pastime. Even though Wells was nominally a pacifist, he extolled the virtues of imaginary battles in which toy soldiers took the place of soldiers' bodies.

The horrors of World War I meant that tabletop war gaming of the kind codified by Wells remained a niche pastime until the 1960s, when the Avalon Hill company began producing strategy games. The Avalon Hill publication *The General* helped establish communities of gamers. The key figure in this process is Gary Gygax, who wrote a "Preface" to Wells's *Little Wars* crediting him as a primary influence. In the context of protests against the Vietnam War, Gygax added fantasy figures taken from J.R.R. Tolkien that distanced battles in Dungeons and Dragons and other fantasy role-playing games from actual war and allowed him and others to claim that they enacted only imaginary violence. While most studies of video games and violence focus on first-person shooters, I analyze turn-taking strategy games that are the indebted to Wells and Gygax. These games reenact the kind of implicit colonialism found in Wells's texts and inculcate a neoliberal subjectivity through in-game mechanics. I explain in the conclusion to the chapter why I play these games myself, even though, like Wells and Gygax, I consider myself a pacifist. I discuss the pleasure in playing them for gamers such as myself and my own involvement in the history of making war into entertainment. Pleasure is omitted from most academic studies of video games, but it is the primary motive for playing such games; however, I analyze how playing these games constructs a subjectivity aligned with wider power structures and encourages a war culture. Playing these games is therefore a guilty pleasure.

Where Wells used toy soldiers, video games replace the body of the soldier with a virtual one as part of the derealization of warfare in the late twentieth and early twenty-first centuries. United by the screen as interface, video games, movies, newscasts, and online videos have contributed to the creation of the "military industrial media entertainment complex (MIME)" (Der Derian xi; Engberg-Pederson 156; Simons and Lucaites 3), which has helped the United States in particular "become fully assimilated to a war culture" (Simons and Lucaites 3). Building on this analysis (in chapter 5), I place images by Sophie

Ristelhueber in the context of what has been termed "drone vision" to show how her photographs of post-battle landscapes without bodies inadvertently reinforce the image of war without bodies. Bodies have been excised from media coverage, making accounts of war primarily about material damage.

This suppression of the body in the American media is reinforced by an emphasis on the aftermath of war as trauma for soldiers, especially in the diagnosis of PTSD (chapter 4). Trauma as a "wound of the mind" has replaced the representation of the damage to bodies by news media in the United States. Simons and Lucaites have examined the trope of "the return" in narratives about American soldiers trying to integrate back into civilian life and see such accounts in terms of "the invisibility of combat trauma" (48); in such narratives we see "not trauma of war but trauma of returning from war" (49). Media coverage voluntarily eschewed showing damage to bodies from combat, and focused on PTSD, substituting therapy and the promise of reintegration for a discussion of the long-term damage from the Gulf and Iraq wars. Just as the figure of the wounded soldier became the vehicle to come to terms with the ambiguous outcome of the Crimean War, the treatment of PTSD became a way of working through the loss of life in these wars. Rather than an accounting of the cost of an invasion that resulted only in ecological damage and the loss of life on both sides, especially civilian casualties, concern focused on the psychological effects of combat. In this chapter I am not questioning the need for therapy for veterans following trauma of war but rather, like Didier Fassin and Richard Rechtman in *The Empire of Trauma: An Inquiry into the Condition of Victimhood* (2009), critiquing the political use of PTSD as a category (see also Young).

In my conclusion, I address the military dream of an automated "war without bodies," in which violence is promulgated at a distance. Operators of drones thousands of miles away from their target view the terrain and unleash weapons via a video screen, making death into pixelated images and therefore closer to a video game than combat (Derek Gregory 188–189). Automated supersoldiers, which are at the moment the stuff of science fiction like the *Terminator* films, are another way in which the military is moving toward war without the involvement of bodies. The overall aim is to remove precarity for the military by taking the vulnerable human body out of the conflict. This promises to exacerbate the trend that started with wars in the twentieth century where civilian casualties outnumbered those in the military. As I argue in the conclusion, representations of a "war without bodies" make death acceptable to the very people who will suffer most from the effects of violence: civilian populations. We are not necessarily moving toward war totally without bodies, but rather war without soldiers' bodies, in which the "collateral damage" is primarily to the civilian population.

Media coverage of war does not reflect the historical increase in civilian casualties, and my analysis parallels the work of artists such as James Bridle's *Dronestagram* (2012) and Gohar Dashti's *Today's Life and War* (2008) (see chapter 5), which reinsert the civilian body into images of military violence to challenge the "dominant discourses on power and war" (Alison Williams "Disrupting" 12). Their images "break the frame" around conventional discourse about war to recuperate the lives excluded from recognition as casualties. Their aim, as is mine, is to make people aware that "war without bodies" does not mean antiseptic violence without casualties, but rather the sanitization of violence that excludes the deaths of civilians, who have become the primary victims of warfare (see UNICEF).

CHAPTER 1

Sacrificial Bodies

FENTON, TENNYSON, AND THE CHARGE OF THE LIGHT BRIGADE

Previous feminist critiques have unpacked the gender-based hegemony that made Victorian women subordinate their desires to domestic duties, and a similar dynamic was at work for men and war.[1] In a contemporary refutation of Victorian gender roles, John Stuart Mill in *The Subjugation of Women* asserted that Victorian men wanted "not a forced slave but a willing one" and that "all women are brought up from the very earliest years in the belief that their ideal of character is the very opposite to that of men; not self will, and government by self-control, but submission, and yielding to the control of other" (Mill). The masculine equivalent of women as "willing slaves" in this period was the ideology that men were born to sacrifice themselves on the battlefield as part of their biological destiny. The imperative of "duty" linked gender ideologies in household conduct manuals aimed at women, and poetry and imagery that featured men's role in war.[2] Where "duty" confined women to the domestic sphere, the word for men connoted the ultimate self-sacrifice of death in battle. The classic example of such male self-sacrifice was the charge of the Light Brigade during the Crimean War, an event that was described in prose dispatches by war correspondents such as William Howard Russell and in an outpouring of commemorative poetry, most famously by Alfred, Lord Tennyson. These accounts are unified by their invocation of the male sacrificial body as redemption for the folly of British cavalry charging directly at Russian artillery, making a suicidal attack into a heroic self-sacrifice in the name of duty.

The ideology of domesticity figured women "collectively as the feminine essence of the nation" (Poon 22), in parallel with the discourse of military masculinity that represented the dead male body as reaffirming national values.[3] Graham Dawson argues that "within nationalist discourse, martial masculinity was complemented by a vision of domestic femininity" (2), making both genders in their separate spheres exemplars of British identity. The most widely cited statement on Victorian gender roles, John Ruskin's "Sesame and Lilies," also encoded masculinity as destined for battle as a counterpart to female domesticity; woman's role is "not for battle," but the man's role is "the doer, the creator, the discoverer, the defender," and he is destined "for war, and for conquest, wherever war is just" ("Sesame" 122–123). Ruskin's formulation was extreme in its depiction of such diametrically opposed gender roles. Recent analyses have shown that men participated in the domestic sphere and expressed emotions that they might not display in public. For instance, Holly Furneaux's *Military Men of Feeling: Emotion, Touch, and Masculinity in the Crimean War* (2016) has directly countered the image of an unemotional British military masculinity. Furneaux argues that "many soldiers made strenuous practical and imaginative efforts to connect themselves with the structures of home life" (13), extending the critique of "separate spheres" into the military.

This is an important corrective to a stereotype of Victorian masculine identity, but I am concerned in this chapter with the representation of soldiers in photographs and texts that align themselves with dominant Victorian gender norms encoded in language. Kate Flint argued in *The Woman Reader, 1837–1914* (1993) that, while Victorian women may not have followed the strictures of conduct manuals in private, the "reiteration served as a confirmation and consolidation of the dominant ideology" (116–117); the same argument can be made for Victorian masculinity, because language reinforced certain roles for men and prohibited others, at least in public. Florence Nightingale in her prescient analysis of Victorian gender roles, *Cassandra*, analyzed the hegemony of such norms, noting terms of opprobrium for men who strayed too far into the domestic sphere, such as "knights of the carpet" and "drawing-room heroes" (32). The terms encode an image of warlike masculinity in "knight" or "hero" that was seen as inimical to the domestic, evoked by words such as "carpet" and "drawing-room." Men were supposed to be in the public sphere rather than at home, and in metaphorical and sometimes literal combat. Tai-Chun Ho has documented how noncombatants writing about the Crimean War were satirized as "pianoforte poets," derided for describing battles that they had not witnessed while "at home at ease" (509). The reference to the pianoforte deliberately subverts the authority of the poets by locating them in the home rather than on the battlefields of the Crimean War. Such criticisms reinforced a public image of aggressive, heroic, military masculinity that was defined as out of place in the domestic sphere.

According to John Tosh, Victorian masculinity was "premised on a powerful sense of the feminine 'other,' with each sex being defined by negative stereotypes of the other" (91). In Ruskin's formulation, men are predisposed to war, so that dying in battle confirms their manliness, making military masculinity the symbolic epitome of Victorian manliness.[4] In dying, the men in the charge of the Light Brigade thus affirmed their duty to sacrifice the male body in war,

Figure 1. Roger Fenton, *Valley of the Shadow of Death*. Courtesy of the Getty Centre.

thereby fulfilling their gender role in a heroic final act. They exemplified the "good death" trope in which the greatest heroism was attributed to those killed in action, especially the self-sacrificing deaths of those who "'offered their bodies to save particular friends'" (Furneaux 130–131); in the hierarchy of military death, such willing self-sacrifice was the preeminent example of heroism.

The Crimean War was notable for the effect of technology on print and visual media. Roger Fenton took some of the first photographs of the British army during a campaign, and Russell's dispatches were transmitted to newspapers and their readers by telegraph.[5] Trudi Tate refers to it as "the first modern war" because of these new technologies (162). Tate also notes that the "pianoforte poets" were dependent on Russell's dispatches because they did not witness the events themselves (178). However, this novel representation of war helped connect patriotism and nationalism through coded language and visual imagery, and was assimilated into existing paradigms rather than fundamentally changing perceptions of the British military. Fenton's photographs, Russell's dispatches, and Tennyson's poetry framed the deaths of soldiers in such a way as to reaffirm rather than question the justifications of such sacrifice implicit in Victorian representations of war. The most iconic image of the Crimean War, Fenton's *Valley of the Shadow of Death*, obliquely evokes these justifications while documenting warfare without showing any dead bodies.

Documenting the Crimean War: Fenton's Photographs

Fenton provided some of the most memorable images of the Crimean War, but the contrast between his diary entries and the subjects of his photographs underscores the implicit ideological frame that constrained him; this is not to say that he was censored, but that in eschewing images that would undermine a heroic image of the British military, he operated within a set of conventions that emphasized masculine self-sacrifice.[6] As Jennifer Green-Lewis notes, "to one looking through the camera, of course, the world is instantly and already framed" (123), and this is true both literally and ideologically of Fenton's images. He produced photographs that would be acceptable both to the military and to a Victorian civilian audience because they were located within a discourse of heroic, self-sacrificing masculinity.[7] His photographs therefore did not challenge the dominant narrative of the British armed forces as a powerful fighting force, but rather reinforced existing attitudes to war. There is also a marked contrast between the death he witnessed while he documented the British army at war and what he represented in his photographs. On June 2, 1855, Fenton retraced the route previously taken by the ill-fated Light Brigade on the October 25, 1854:

> On our way we went exactly in the line taken by our cavalry at Balaclava, except that we met instead of following their line of advance. We came across many skeletons half buried, one was lying as if he had raised himself upon his elbow, the bare skull sticking up with still enough flesh left in the muscles to prevent it falling from the shoulders; another man's feet and hands were out of the ground, the shoes on his feet, and the flesh gone. (Gernsheim and Gernsheim 87)

Fenton relived the charge of the Light Brigade and saw its aftermath, but he did not make a photographic record of the skeletons, nor did he make any overt commentary on the military blunder that led to such carnage.[8] Fenton also did not record any particular emotional reaction to the grisly scene, in contrast to spectators such as Frances Duberly, who described trembling and shaking upon seeing dead British, Russian, and Turkish soldiers after the Battle of Balaclava (80), or Russell, who reacted in shock when witnessing "the horrors of the battlefield" (52), where "one could not walk for bodies" (53). Fenton took many photographs while in Crimea in 1855, but none of them depict the dead, although he clearly saw many examples of the gruesome aftermath of such battles. Above all, Fenton's photographs were consistent with a class-based aesthetic that represented the officer class and the cavalry as the epitome of British masculinity and reinforced an image of war as a noble spectacle.

Fenton's class status is signaled by the next line in his diary after this entry, where he mentions in passing that "when I got to Headquarters I found Lord Raglan's invitation to dinner" (Gernsheim and Gernsheim 87). Fenton was well connected with the upper echelons of the British army and dined regularly with officers. He also used the promise of portraits to gain favors from officers when he needed help transporting his mobile developing studio around the Crimean Peninsula. His studio was a converted butcher's cart, and he often needed help navigating the rutted Crimean dirt roads, and the bribe of a portrait would entice officers to lend their men's muscles to the effort, although it also meant that he would be pestered when in his studio trying to work by other officers knocking on his door asking for their portraits to be taken.

Armies move in large numbers, so that Tennyson referred to the Light Brigade as the "six hundred," for example, and William Simpson's watercolor for *The Seat of War in the East* (1855–1856) depicting the charge shows indistinct masses of cavalry in straight lines. Fenton's photographs of cavalry, such as the portrait of Cornet Henry John Wilkin, individualized the war effort and presented dashing images of the officer class.[9] As Joan Hichberger says, "Fenton directed his lens toward agreeable aspects of the war," including "officers looking cool and competent in smart uniforms" (51). In focusing on a mounted officer such as Wilkin, Fenton also reinforced "the collective perception . . . that virility and an aptitude for riding are linked" (Laneyrie-Dagen 131). When photographing groups such as councils of generals and officers, Fenton presented similarly forceful images so that the "incompetence of Lord Raglan's leadership . . . which had been widely criticized, was eclipsed by his portrait" (Green-Lewis 101), which presented him as a dynamic leader. His images thus omitted the incompetence of the British military's upper echelons. Unlike Russell's dispatches, Fenton's photographs also excluded the suffering of ordinary soldiers.

The *Punch* cartoon "Grand Military Spectacle: The Heroes of the Crimea Inspecting the Field-Marshals" (November 3, 1855), by contrast, shows injured enlisted men faced with the finery of the officers, implicitly calling into question their competence and lack of concern for the average soldier. Fenton's photographs such as *Lieut.-General Sir George Brown and Staff* (Gernsheim and Gernsheim 89) shows the commanding officer as the center of the organization, given that one of the staff at whom he is looking seems to be briefing him on the state of the campaign while the rest look on attentively. The photograph conveys an image of military command and control. Another photograph, entitled *Cookhouse of the 8th Hussars* (Gernsheim and Gernsheim 60) depicts a group of soldiers who participated in the charge but whose unit was out of action by the time it was taken because the number of casualties meant that they were no longer able to fight (Lalumia 119). The scene shows a moment of apparent leisure, but without reference to the casualties inflicted in the ill-advised charge. Fenton's photographs thus omit suffering and death from the frame. Fenton's photographs systematically exclude the injured or dead bodies found in *Punch* cartoons or images in the *Illustrated London News*.

Figure 2. Roger Fenton, "Cornet Henry John Wilkin." Courtesy of U.S. Library of Congress, Prints & Photographs online catalogue, http://loc.gov/pictures/resource/cph.3g09124/.

Fenton is often credited with being the first war photographer, but that label is erroneous, because he would more accurately be described as the first photographer to document what the British army looked like when it was fighting a war.[10] Fenton could not photograph live action due to the limitations of the collodion wet-plate process; therefore, his images are either humans standing still or the aftermath of battles;[11] he could, however, have photographed dead bodies if he had wished, and there has been much debate over why he avoided images of corpses (Lalumia 118–119; Ulrich Keller 123; Groth 557). There was no cultural prohibition against photographing dead bodies per se, as attested to by the convention of dressing dead loved ones up as if alive and posing them for funerary portraits.[12] The deceased were, of course, ideal subjects in that they would not move during the long exposure times needed for early photographic processes and were posed in postmortem photography as if alive.[13] Despite the convention of mourning portraits of dead loved ones, Fenton did not seem to regard dead soldiers as a possible subject; his photographs are more a sort of "virtual tourism" of the Crimean landscape (Groth 553), and a flattering portrait of various officers, rather than a documentary of the violence of war, which must be inferred from the destruction left behind. Indeed, his most famous photograph is simply of cannonballs lying in a valley.

The image is powerful both because of what it omits and because of its title, which echoes Tennyson's poem on the charge and its refrain "into the valley of death." There are no bodies visible, living or dead, only an empty landscape as a mute testament to war. Despite the association with Tennyson's poem through its title, this is not in fact a photograph of the valley down which the Light Brigade charged; it lay behind a British redoubt, and when cannonballs missed their target they landed in this valley, thus littering the landscape. Also, the title was appended by the authors of the catalogue of his exhibition in the autumn of 1855, not by Fenton himself; nevertheless, as Groth emphasizes, none of this "detracts from the tenability of the connection between Fenton's image and Tennyson's poem" (553). Fenton and other photographers were well aware of what Groth terms the "literary grid" (554) (her term for the frame) that informed their aesthetic and the links between their images and other representations of the Crimean War.

Ulrich Keller has argued that in the Crimean War "aesthetic factors enter the analysis at every point" (4), and this is true of Fenton's *Valley of the Shadow of Death*. The photograph is a landscape with still life, with some critics raising questions as to whether Fenton rearranged the cannonballs himself for artistic effect.[14] The image gains resonance not by showing the effects of war directly, but rather by allowing an imaginative reconstruction through its depiction of the aftermath of a battle. There are no human figures in this landscape, unlike a similar photograph of the Victoria Ravine, with soldiers collecting cannonballs from the road (Keller 134). The different angle and the inclusion of British soldiers in this photograph by James Robertson makes for a far more prosaic image shot from a distance, whereas Fenton's photograph focuses attention on the cannonballs, with nothing to distract the viewer and nothing to limit the imagination. Fenton's image is indirectly elegiac, but the title also carries with it an implicit religious consolation by its reference to the "valley of the shadow of death." The title recalls Psalm 23:4 ("Even though I walk through the valley of the shadow of death, I will fear no evil, for you are with me; your rod and your staff, they comfort me"), so that death is a gateway to religious salvation. Viewing a landscape such as this through biblical references was part of the tradition in both art and photography (Weaver 106); thus, the religious references in Fenton's landscape would have seemed natural to a Victorian audience and would have tempered the sorrow at the loss of life.

Fenton's aesthetic approach was echoed in Russell's dispatches. He set the stage for his report on the Battle of Balaclava by looking at the landscape with a "painter's eye" and extolling "the beautiful scene" (66), obviously in part as a bucolic contrast to the violence to come. However, also he had frequent recourse to aestheticizing vocabulary, describing skirmishers as "like moonlight on the water" (67) and the cavalry as having "a halo of flashing steel above their heads" (73). Such vocabulary framed war within the discourse of painting and located the violence of the Battle of Balaclava within an artistic genre that allowed appreciation of the beautiful aspects of warfare. Russell was not deliberately trying to make warfare palatable, but his aesthetic appreciation of the scene before him helped sanitize the violence for his readers. The same normalizing effect on the representation of war governs Fenton's photograph of the valley by mixing landscape and indirect mourning in its invocation of the "valley of the shadow of death."

Fenton's image, thanks to its title and association with Tennyson's poem, functions as a virtual war memorial through which the spectator can imagine the death of soldiers while finding consolation by the way it is framed. The photograph is a site of mourning that mixes nationalism, the celebration of heroic masculinity, and a remembrance of the dead in much the same way as poems written about the charge of the Light Brigade or the Crimean War monument in Waterloo Place, London (see chapter 2). The Crimean War monument in Waterloo Place is a physical "site of memory," whereas Fenton's is a virtual space, but both mix death and heroism in the same way.[15] The bodies of the soldiers in the memorial both invoke military masculinity and recall the deaths incurred in the conflict as an example of a patriotic "*mort pour la patrie*" (death for one's country) in Antoine Prost's typology of memorials (312). National pride in "*mort pour la patrie*" serves the same function as religion in *Valley of the Shadow of Death* by invoking a justification for war. Poetry about the charge of the Light Brigade carries out the same ideological work by justifying war as noble, masculine self-sacrifice.

Reliving the Charge of the Light Brigade

Poetry memorializing the charge of the Light Brigade framed death in a way that recuperated the loss of life as shared sacrifice in the name of duty.[16] As Cynthia Dereli says, the function of poetry about the Crimean War was to "develop common images of shared, though still painful, emotion and to counsel resignation" (120). Fenton's photograph and Tennyson's poem proffer religious consolation for the self-sacrifice of the Light Brigade. Along with the nationalism of the poem, the religious context gives meaning to what otherwise would be a senseless waste of human life. The Crimean War itself was justified partly on religious grounds (Figes 5), and it became a nationalist cause once Russia was identified as the enemy. Such consolation helped blunt the force of the criticism of the charge as a military blunder and to turn Tennyson's poem into a memorial rather than a protest. Jerome McGann's comment that Tennyson's poem "is not so much a commentary on the war and British foreign policy as it is a eulogy of the British character" (191) captures well the dual function of the poem: mourning the deaths of the cavalry while reaffirming British military masculinity.

Numerous words connote action in Tennyson's poem, endowing it with a sense of the cavalry charge as a piece of dynamic and exciting bravery. As Jason Camlot says, the "dactylic metre of the poem seems to invite a fast and furious reading"; Camlot goes on to quote R. M. Milnes's observation that the "charge" is "a real gallop in verse, and only good as such" (27). The poem uses words like "charged," "plunging," "rode," and "forward" to convey the cavalry charge as a headlong rush, whereas eyewitness accounts emphasized the measured and orderly way that the Light Brigade approached the artillery until they came under fire (Tate 167). Tennyson's is a civilian image of a cavalry charge as a thrilling gallop and the epitome of heroic British military masculinity. The poem is a visual analogue to paintings of the Light Brigade such as that by Thomas Jones Barker, *Charge of the Light Brigade* (1877), which shows the cavalry slashing at hapless Russian artillerymen as they gallop through their ranks with sabers raised.[17] Tennyson's poem thus aligns with a more widespread celebration of Victorian military masculinity and death in battle.[18]

The use of "rode," "charging," and "plunging" also alludes to the horse as an integral part of the scene; indeed, the cavalry could be referred to as "horsemen," thereby making them a composite of human and animal (Waddington 29). In poems about the charge of the Light Brigade, the horse is often represented as an extension of the soldier's own lust for violence. In *Duty: or, the Heroes of the Charge in the Valley of Balaclava*, the horses also stand for the belligerence of the nation as they await the start of the battle. They become "Our own brave British horse":

We hear the champing of the bit—
We hear the war-horse neigh—
We see our men in saddle sit,
And heroes then looked they (Waddington 59)

The cavalryman on a brave "war-horse" is a hero thanks to the connotations of the animal with battle, honor, and images of the armored "gallant knight" (Kestner 98). The attitude to horses in these poems contrasts markedly with that of Frances Duberly, whose most important relationship, apart from that with her husband, was with her horse Bob; she reacted with equal horror to the death of humans and horses in battle, whereas in poetry such as this the animals are seen as eager participants. The ascription of warlike feelings to the horse as well as the soldier reaches a somewhat absurd apotheosis in Henry Sewell Stokes's "Balaklava," where the charger reaches positively supernatural heights of aggression:

Hark! The trumpet sounds from far,
Quick the charger scents the war
Either eye-ball like a star
At the aspect bright'ning;
How he paws the mountain-sod,
Treading as if air he trod
While his hoofs seem thunder-shod
Wreathed his man like lightning! (Waddington 68)

An obvious inaccuracy in the poem is that the Light Brigade weren't on a mountain but in a valley, but the poem is not aiming at realism. The horse, which can apparently levitate and send lightning from its hoofs, is meant in this passage to endow "his man" with more than human strength and power. Like the saber, the horse represents the masculine violence and power about to be unleashed in the charge against the Russian artillery. Such overexcited verse shows the effective understatement in Tennyson's "Charge of the Light Brigade," which simply states that "horse and hero fell." Tennyson elides the fate of the horse and rider as well, but he does not attempt to endow the animal with human qualities or encumber the animal with elaborate similes.

Tennyson's other, less well-known poem about another cavalry charge during the battle at Balaclava, gives some sense of why his minimalist account of the Light Brigade action was so memorable. "The Charge of the Heavy Brigade" does not have same rhythm or the sense of excitement of his earlier poem, and often loops ponderously back on itself to ask rhetorical questions that interrupt the narrative, such as "and who shall escape if they close?" which is spoiled if you know the outcome of the encounter and that the answer is "almost all of them" (Tennyson 373). The reader knows of the impending doom of the six hundred, so questions that in "The Charge of the Light Brigade" had pathos fall flat in a poem about a successful battle for the British against Russian cavalry; it also comes across as forced when it tries to convey the charge through lines like "The trumpet, the gallop, the charge, and the might of the fight!" which substitutes a list for the more evocative opening of "The Charge of the Light Brigade," which begins with the charge already underway:

Half a league, half a league,
Half a league onward,
All in the valley of Death
Rode the six hundred. (Tennyson 147)

As Ruskin said in his accurate but rather chilling remark, the "soldier's trade, verily and essentially, is not slaying, but being slain" (*Unto This Last* 33), and the poem has the Light Brigade as already galloping to their doom in the "valley of Death." Tennyson's poem, like Fenton's image, is elegiac, suggesting that the Light Brigade was charging to its death, whereas the Heavy Brigade triumphed. While it is true the British cavalry were greatly outnumbered, the brief combat did not produce many casualties because of blunt sabers and heavy Russian overcoats (Figes 246; Spilsbury 152). The charge of the Heavy Brigade does not ultimately have the same resonance as that of the Light Brigade, because it does not represent the action as an example of heroic self-sacrifice. The difference shows that impending death was an essential component of the representation of British military masculinity in "The Charge of the Light Brigade."

While Tennyson's poem is elegiac, a *Punch* cartoon showing a man reading an account of the charge in a newspaper shows the mixed emotions the charge generated (Leech 213). The man holds a poker above his head in imitation of a saber, and in high excitement as he reads the account to his assembled family. Entitled "Enthusiasm of Paterfamilias on Reading the Report of the Grand Charge of British Cavalry on the 25th," it satirizes the overenthusiastic man and his imaginative reenactment of the action as he reads the newspaper account of the charge of the Light Brigade. The man's excitement is conveyed by the poker held over his head to simulate the "flashing sabres" wielded by the cavalry. In contrast, the mother holds her head in her hand while pressing a handkerchief to her forehead, and a young woman looks on with a distraught expression (see Markovits 186–192). In the image the women convey the grief that accompanies the death of the cavalry, while the man represents lines such as "Flash'd all their sabres bare/Flash'd as they turn'd in air" (Tennyson 148), identifying with the cavalry who reached the Russian artillery and sabered them. As Tate argues, the "*fantasy* investment in war" in this image is divided along gender lines (Tate 165; italics in original), with the man representing "enthusiasm" and a vicarious excitement, and the women representing mourning. The man conveys his investment in an image of heroic military masculinity that prefigures H. G. Wells imagining himself as a general or that of a player in a video game commanding virtual troops (see chapter 3). The cartoon also underscores the status of war as enthralling spectacle for those who do not have a direct role in the violence; they encounter war through simulations that make war exciting and thus enjoyable. Where those who wrote verse about the war were "pianoforte poets" (Ho 509), the man is a "pianoforte soldier" imagining himself in the place of the cavalry charging the Russian artillery.

The use of "flash" in Tennyson's poem indicates how much of the allure of a cavalry charge was visual, as does Russell's description of the "glittering masses" of British cavalry (67). McGann says of this imagery that it makes the Light Brigade into "an aesthetic object" and links the poem to a tradition of "heroic military art" (199); to this I would add that it continues a tradition of representing military masculinity in the same way as Fenton's photographic portraits of cavalry officers. Furthermore, the cavalry saber is a motif that is found throughout poems about the charge and functions as a symbol of aggressive military masculinity. The phallic connotations of the saber are obvious and have been well documented in British military imagery by Joseph A. Kestner in *Masculinities in Victorian Painting* (1995).[19] Most striking is the way in which the male body and the weapon are fused; the saber is represented as an extension of the male body and, like the uniform, transforms the cavalry into a spectacle.[20] The saber is a prosthesis fused with the biological body as an integral part of masculine prowess and violence.

The visual appeal of light flashing from the surface of the blade indicates how war is framed as an aesthetic experience in poetry about the charge, just as Fenton's photographs treated the Crimean valley as an aesthetic setting. Appreciation of war as a performance was reinforced because "military events were methodically organized as spectacles" (Keller 4) and made visually appealing.[21] The motif of the flashing saber in poetry also turns war into spectacle and is found throughout the poems collected by Waddington in numerous lines.[22] Russell too describes how the Light Brigade's sabers could be seen "flashing through the smoke" (73). The "flashing" turns combat into a glittering spectacle that masks the brutality of warfare. The sabers are "bright" and alluring rather than bloody, and thus aestheticize violence.

The same view of war as an aesthetic object is evident in Russell's descriptions of the charge, which, as much as Fenton's photograph and Tennyson's poem, indicate to the reader that the engagement is to be appreciated in visual terms. An observer of the charge taking place in the valley below, Russell invokes theater as a way of understanding the unfolding scene: "Lord Raglan, all his staff and escort, and groups of officers, the Zouaves, French generals and officers, and bodies of French infantry on the height, were spectators on the scene as though they were looking on the stage from the boxes of a theatre" (Russell 69).

Using the example of the theater interpolates an emotional distance from the action immediately, although Russell's intention here is to underline the helplessness of the officers as the tragedy unfolded before them. The proscenium arch of the theater frames war in the same terms as the frames around Fenton's photographs, and normalizes war as an extension of other civilian activities; the analogy helps make the scene understandable for readers, but also anesthetizes the violence to come. Russell describes the charge as a "fearful spectacle," but such language already turns it into an aesthetic object to be appreciated like a *tableau vivant*. In an editorial, the *Times of London* wrote to its readers that "to us war is a spectacle," and they could find it "amusing" if no relatives were involved ("Every Man" 6), as it was for the man waving a poker while reading the newspaper in the *Punch* cartoon. While Russell tries to convey the horror with which he and others witnessed the charge, he also echoes Tennyson's poem in shaping how the reader should react emotionally, especially in his invocations of heroism:

> They swept proudly past, glittering in the morning sun in all the pride and splendour of war. We could hardly believe the evidence of our senses. Surely that handful of men were not going to charge an army in position? Alas! It was but too true—their desperate valour knew no bounds, and far indeed was it removed from its so-called better part—discretion. They advanced in two lines, quickening the pace as they closed towards the enemy. A more fearful spectacle was never witnessed than by those who, without the power to aid, beheld their heroic countrymen rushing to the arms of sudden death. (73)

Like Fenton's photographs and Tennyson's verse, Russell's references to nobility and honor describe a heroic masculinity that is achieving its destiny in the "theater" of war. Russell frames the response to the passage by using the words "proud," "pride," and "valour" in connection with the charge, emphasizing the cavalry's stature and heroism as emblems of British military masculinity. The word "valour" connotes heroic self-sacrifice as a natural part of military heroism despite the obviously suicidal nature of the charge. In a rather chilling formulation, Joseph Bristow asserts that these "blunders, errors and disasters committed in the name of one's country" confirm national identity and "make men into *men*" (Bristow 143; italics in original). The adjectives "desperate" and "fearful" recognize the tragedy unfolding, but like "theirs not to reason why" in Tennyson's poem, the horror at the senseless loss of life is overwhelmed by the emphasis on heroism and obedience to the cavalry's duty to obey orders. As in Tennyson's use of "blunder," there is a hint of criticism in the reference to the lack of "discretion," but it is a weak word in comparison to possible descriptions such as "foolhardy" or "suicidal."

Lloyd Demause's survey of the rationales given for war identifies the most frequently voiced sentiment as "suicidal, sacrificial motivations" (232), vocabulary that rationalizes the death of soldiers as self-sacrifice. Similarly, François Lagrange has analyzed the "radical vision of self-sacrifice in war leading to a valorisation of 'certain death'"[23] in narratives from World War I (63). "Theirs not to reason why" in Tennyson's poem validates self-sacrifice in the face of certain death in the same way as World War I poetry; this rationalization is reinforced by "All in the valley of Death," which emphasizes the cavalry as entering a killing field. This idea is specific to men and reinforces the idea that their highest form of duty is self-sacrifice in battle. A contemporary review of "The Charge of the Light Brigade" describes the "noble obedience" of the six hundred and then goes on to say that "the crowning virtue of the dead is declared to have been the virtue of obedience, that of self-subjugation to the law of duty" (Dowden 331). Obedience unto death affirms a set of values that extol self-subjugation to the ideals of queen and country no matter what the cost.[24]

Tennyson's poem and other poems from the same period about the charge of the Light Brigade invoke "duty" as the ideal that unites the civilian reader with the military. Duty "is a function of death" in Tennyson's poetry (McGhee 66). Tennyson's line "Theirs not to reason why" indirectly references duty as obedience. Reviewing the poem, Dowden wrote that "Mr. Tennyson conceives the virtue of noble obedience" in the poem and later extolled "the virtue of obedience, that of self-subjugation to the law of duty" (331). The unquestioning obedience of the Light Brigade in following an order that they knew to be folly is cited as the epitome of duty as obedience to authority; one might be tempted to say blind obedience in this case, but in Victorian terms being willing to sacrifice one's life in this way was an exemplary action that embodied duty in its highest form. Poems on the charge frequently link heroism and duty, as in Caroline Hayward's "The Battle of Balaclava, and the Unparalleled Heroic, but Deeply to Be Lamented

Charge of the Light Cavalry, through a Mistaken Order," which in its elaborate title signals the way in which the poem recuperates the folly of cavalry charging directly at artillery:

> Duty! The magic of that name,
> In British breasts lights valour's flame
> They onward speed, though each man knew
> 'Twas certain death he speeded to
> Proudly they swept, in glittering pride. (Waddington 111)

The "magic" of duty rescues riding to certain death from being a tragic waste of lives because of the "mistaken order" into an occasion for national pride. Rather than a monumental military blunder, the action is transformed into a shining example of self-subjugation on the part of the cavalry; not only do they willingly sacrifice themselves, but they ride "in glittering pride," a description that evokes the same aesthetic appreciations as the account by Russell. Like Tennyson, Hayward turns the charge into a "noble" act that evokes wonder at masculine self-sacrifice in battle:

> Oh, noble heroes of that day
> By those, who prompt at *duty's* call,
> Rejoicing for their country fall! (Waddington 112; italics in Hayward's original)

Hayward reinforces the importance of duty in justifying the charge by putting the word in italics, as if it were not already obvious that this is a central justification for elevating it to heroic status. It strains the imagination (and contradicts the firsthand accounts of what being in the charge was actually like) to say that members of the Light Brigade were "rejoicing" at their deaths. "Rejoicing" is the reaction that Hayward and other poets hope to elicit by framing the charge within the multiple associations of "duty," turning the horror of their sacrifice into an occasion of national celebration. This is one of the more extreme examples of how violence is aestheticized in such poetry. The actual reaction of the members of the Light Brigade was sober even though they knew the charge was a mistake. The need to obey an order that one of them called "perfect madness" (Dawson 139) was expressed as their "duty":

> I could see what would be the result of it, and so could all of us; but of course, as we had got the order, it was our duty to obey. I do not wish to boast too much; but I can safely say that there was not a man in the Light Brigade that day but what did his duty to his Queen and Country. It was a fearful sight to see men and horses falling on all sides. (Dawson 153)

Duty compelled members of the Light Brigade to obey an order that they knew was mistaken "for Queen and Country." Even a writer like Tom Taylor in *Punch*, who, unlike Tennyson and Hayward, was explicit in calling for blame to be ascribed to the person who gave the order, linked duty, heroism, and death:

> But for your true band of heroes, you have done your duty well:
> Your country asks not, to what end; it knows but how you fell! (Waddington 32)

In these lines Taylor retreats from "blame" to rejecting the question of what the charge accomplished in favor of praising the action in terms of "duty" and death. Earlier in the poem Taylor also linked the charge and masculinity, writing that "the order came to charge, and charge they did—like men" (32), reinforcing the association of heroic military masculinity with inevitable death. This sentiment finds its purest, most absurd expression in John Westland Marston's *The Death-Ride: A Tale of the Light-Brigade* (1855), where a capitalized DUTY speaks and "only said, 'Die'—and they died" (Waddington 41). Both Taylor and Hayward also read the charge in nationalist terms, representing the British being united as a country through "duty." Duty thus links heroism and nationalism and allows a vicarious civilian identification with the Light Brigade through a shared British identity experienced as a collective imaginary based on self-subjugation.

"Duty" linked citizens with the government in that "the duty of a State [is] to inculcate religion and morality among the great body of the people" (from *Hansard*, quoted in Parry 94), and the duty of the people was to support the state; in other words, duty was an instrument of governmentality. It was not just that nationalism gave people a way of identifying with the larger project of the Crimean War as a defense of British liberty; as Parry shows, the Liberal government saw itself as engaged in a "warfare of thought" in promoting its values (100). In Clausewitz's terms, this was "politics" making sure that the engine of public opinion drove the machinery of war (see Introduction). The government's rhetoric militarized terms like "liberty" and "duty," and made them extensions of the British Crimean campaign and helped ensure support for the war once it had begun, in a parallel to my description in the Introduction of the propaganda leading up to the war in Iraq. The dead cavalry can therefore be viewed as carrying out a literal war that is also waged on an ideological level as a defense of essential British values.[25] The Light Brigade was seen as affirming these essential British values in death, just as conduct manuals like Sarah Stickney Ellis's *The Women of England* (1839) extolled women as exemplars of domestic probity in subjugating their wills to the demands of household and family.

"Duty" was therefore a term that linked the subject and the nation for both genders. Duty was both internal, in that self-subjugation exemplified the moral rectitude of the British, and an imperative from the nation and the queen that must be obeyed no matter what the cost. Through vicarious identification with the charge of the Light Brigade, Victorian readers could affirm their own sense of identity and their participation in the imaginary collective of the nation by seeing it as a product of a uniquely British code of conduct (which extended also to the British cavalry's horses). The charge of the Light Brigade was understood through the ideology of "duty," not as a military blunder, but as the epitome of masculine self-sacrifice, a sacrifice that affirmed the superiority of British cultural values.

The cavalry in Fenton's photograph and Tennyson's poem were, in J. A. Mangan and C. McKenzie's terms, "sacrificial warriors" who reinforced "an ideal of selfless service to the state," as well as "engender[ing] uncritical conformity to the values of the group" (1101). They also served to delineate the contours of British masculine dominance in contrast to other nations and ethnic groups as part of "the need to establish a sense of racial superiority as a cornerstone of this selflessness" (1101). In the earliest known poem about the charge, in the *Manchester Guardian*, the Light Brigade is described as both fighting and dying in a way that proves British national superiority, as "they fought as Britons only fight, and fell as Britons fall" (Waddington 29). The bravery of British cavalry was frequently contrasted with Turks "like cowards dying!" (Waddington 31), as nationalist stereotypes were used to disparage an allied army, although more objective historical accounts show the caricature to be unjustified. The Turks were vilified, and the allied French largely ignored, despite the Chasseurs d'Afrique coming to the aid of the Light Brigade, but they do get an honorable mention in J. N. Cranston's *The Fall of Sebastopol* (Waddington 119). Lord Raglan, commander of the British troops, in his account of the charge, noted that "the Chasseurs d'Afrique advanced on our left, and gallantly charged a Russian battery, which checked its fire for a time, and thus rendered the British cavalry essential service" (Grehan and Mace 71), gallantry that was not recorded in most poems on the charge because it would have undermined boasts of British superiority. Both Turkish and French troops were excluded from the frame of "heroism," which was extolled as a uniquely British trait.

However, contradictions in this nationalist ideology emerged in some cases because another index of British superiority was supposedly "freedom."[26] Russian infantry, made up largely of serfs, were disparaged as "Russia's slaves" (Waddington 56) and as an indistinct mass in which "Full thirty thousand men were there—/No!—*thirty thousand slaves*" (Waddington 57; italics in original) blindly obeyed their orders. Unlike the British forces, the Russians were seen as inadequate because they were not "free" and were thus vanquished by British liberty: "The serfs of the despot put to flight/By the strong arm of the free" (Waddington 59). This vaunted British freedom was also paradoxically tied to the nation and the queen, which both demanded obedience, but not servitude: "To land and Queen a freeman's love/Each gallant heart controls" (Waddington 64). The authors seem unaware of the latent contradiction in their verses, when blind obedience was expected and questioning orders was not allowed, but the British army was exalted as champions of British freedom. Holding the military up as the standard bearer of liberty resulted in irreconcilable images of both freedom and blind obedience in poems about the charge of the Light Brigade.

Tennyson's troubling line "theirs but to do and die" (147) underscores how the Light Brigade had no choice but to obey and "do their duty." The military is a bad example to choose as an emblem of "freedom," because soldiers are expected to obey orders without question and, while they may have strong arms, they are not at liberty to determine their own fate; neither are the queen's civilian subjects "free" if they are bound by duty to obey her. The contradictions rest upon a definition of "duty" as voluntary self-subjection, with charging to one's death as the most extreme example of duty in action. The members of the Light Brigade were male "willing slaves" in Mill's terms, in that they voluntarily rode to their deaths, and were thus not so different from Russian troops who also obeyed orders. These contradictions in British subjectivity were memorably parodied in W. S. Gilbert and Arthur Sullivan's *The Pirates of Penzance, or a Slave to Duty* (1879), where the word is used to justify many questionable actions, including piracy. Frederic as a pirate says that he is a "slave to duty," which is a joke, of course, but shows how obedience could be parodied as servitude and loss of will (Alan Robbins 65).[27] Frederic later declaims that "but duty is before all—at any price I will do my duty" (92), even though it goes against his conscience (since it means remaining a pirate). Mabel says, "He has done his duty. I will do mine" (96), which parallels Frederic's declaration so that men and women in the comic opera feel the imperative of obedience to duty equally. Duty in *The Pirates of Penance* leads characters to act in morally questionable ways because it is an imperative that must be obeyed without question. While *The Pirates of Penance* is a comic opera, such blind obedience follows the same logic as the Light Brigade that must "do and die."

The invocation of "duty" therefore allowed both men and women to vicariously reinforce their own self-subjugation in reliving the charge of the Light Brigade. The charge became a moral victory rather than a military blunder when experienced in this way, because it allowed for feelings of national pride and the reaffirmation of a shared subjection to the call of "duty," which could be invoked as a rationale for a host of domestic actions, such as in the opening paragraph of Harriet Martineau's reminiscences, where she uses the word three times in justifying her decision to write about herself in response to "the duty of recording my own experience" (*Autobiography*). John Ruskin, in his inaugural lecture as Slade Professor of Art at Oxford University, began by discussing his duty to give the lecture, then later explicitly linked war and service to the expanding Empire:[28]

> . . . recognising that duty is indeed possible no less in peace than war; and that if we can get men, for little pay, to cast themselves against cannon-mouths for love of England, we may find men also who will plough and sow for her, who will behave kindly and righteously for her, who will bring up their children

> to love her, and who will gladden themselves in the brightness of her glory, more than in all the light of tropic skies. (Ruskin, "Lectures" 8)

The malleable concept of "duty" could therefore cover a range of activities, from domestic service, to governing the colonies, to charging full tilt at artillery. Ruskin's segue from war to other kinds of service shows how, for Victorian readers, their own daily activities could be linked to the heroism of the Light Brigade and experienced as affirmation of their own quotidian participation in fulfilling their obligation to the nation and the empire. Farmers, parents, and righteous people in Ruskin's list were fulfilling the same obligation to the nation as the Light Brigade charging Russian artillery. Duty as an imperative could be experienced as both subordination and free will at the same time as a form of national voluntary bondage.

Tennyson's poem does not explicitly use the word "duty," and his minimalist approach to representing the charge is one cause of its longevity. Tennyson also did not explicitly invoke British national identity either, although his reference to "Cossack and Russian" highlights the contrast of these foreign troops with the British cavalry. The poem focuses on the heroism of the charge, without excessive jingoism or moralizing, referring to the "world" wondering and honoring their action rather than emphasizing its purely nationalist import. It does not seem likely that anybody outside the English-speaking world would experience the same emotion at reading Tennyson's poem, because its effect depends on the context of discourse about the charge as an act of selfless obedience that reinforced national identity. The far less effective and unambiguously nationalistic "The Charge of the Heavy Brigade" explicitly says "like an Englishman" in describing how Scarlett led the charge, and refers to "our own good redcoats" (Tennyson 373).

Representing the charge as an emblem of "duty" kept anger and calls for accountability for incompetent leadership at bay. The word that was conspicuously excluded by this ideological frame was "suicidal." While acknowledging the "blunder," poetry on the Crimean War kept its focus on the heroism of the six hundred as a testament to British character and therefore sustained public support for warfare. Only foreign commentators like General Bosquet voiced the idea that the charge was madness and suicide rather than an act of war; British commentary almost exclusively promoted the awe and "wonder" evoked by Tennyson.

The Charge of the Light Brigade as Sacrifice

Tennyson's line "Theirs but to do and die" emphasizes the inevitability of death for the six hundred but raises implicitly the idea of the charge of the Light Brigade as suicide; in phrasing it as "do and die" rather than "do or die," Markovits notes that "the substitution of conjunction marks the difference between bravery and suicide, between 'ordinary' battles such as the Alma, and the extraordinary charge at Balaklava" (158). The phrase raises the question implicitly whether the six hundred, or any man in the British military, had renounced their prerogative to question an order that was obviously counter to the tenets of war and an act of suicide. General Bosquet's famous remark that "*c'est magnifique mais ce n'est pas la guerre, c'est de la folie*"[29] (Ponting 137) underscores that, while one could admire the heroism, the charge defied all military doctrine and was in fact madness. Henry Clifford, who witnessed the charge, recounts a further discussion with Bosquet about the meaning of what they had witnessed:

> The tears ran down my face, and the din of musketry pouring in their murderous fire on the brave gallant fellows rang in my ears. "Pauvre garçon" said the old French General, patting me on the shoulder. "Je suis vieux, j'ai vu des batailles, mais ceci c'est trop."[30] Then the smoke cleared and I saw hundreds of our poor fellows lying on the ground, the Cossacks and the Russian Cavalry running them through as they lay, with their swords and lances. (Clifford 73)

In contrast to Fenton's photographs and the many paintings and poems about the charge, both Clifford and Bosquet experience only grief at the senseless loss of life, which was hard even for an experienced general to bear, even as they admired the heroism of the Light Brigade.[31] This is the emotion that is most obviously excluded from the frame that surrounds most representations of the charge of the Light Brigade; it is not just that Fenton's photographs excluded dead bodies, or that Russell's dispatches emphasized pride and valor, or that poems about the charge emphasized obedience and duty; the searing grief at the loss of life that is found in texts by Clifford or Frances Duberly is muted, and one has to read the accounts of people who actually witnessed the fatal charge to find records of inconsolable loss.

The effect of the frame surrounding Fenton's *Valley of the Shadow of Death* was not to make the loss of life in the charge of the Light Brigade ungrievable, but to temper the sorrow by making it appear inevitable as a fulfillment of masculine military identity and offer religious and nationalist consolation to make it comprehensible as heroic self-sacrifice. Fenton's photograph avoided disturbing this frame by eschewing documentation of the evidence of death he saw when he followed the path of the Light Brigade down the valley, instead inferring such loss through a still life with cannonballs. He could conceivably have photographed the valley complete with skeletons of cavalrymen and horses, but such a graphic image was unthinkable within the frame of heroic masculinity that informed both his portraits of cavalry and poetry about the charge of the Light Brigade.

However, Bosquet's use of "*folie*" or madness also hints at the charge as something so far outside the realms of the normal as to be incomprehensible in rational terms. The many references to the "madness" of the charge by survivors also positions it as an exceptional event that defies conventional logic. "Madness" and war link Tennyson's "The Charge of the Light Brigade," "Maud," and "Locksley Hall." The apparent bellicosity of the ending of "Maud" in connection with the Crimean War has been the subject of much critical debate, especially in the context of Tennyson's attitudes to war (Vanden Bossche 73–74). For Tennyson, war is a kind of insanity because it lies outside the range of normal human experience, and the charge of the Light Brigade was a prime example of its madness, as it was for the Russians who witnessed the action (Danahay). While Tennyson's vocabulary is relatively subdued in his reference to the "wild" charge of the Light Brigade, both "Maud" and "Locksley Hall" are more explicit in connecting warfare with madness and ultimately a self-sacrificial impulse. Markovits cites Herbert Tucker's characterization of Maud's song of "Death, and Honour that cannot die" (I.177) as "traditional" and part of an "aristocratic past" (149), but the song invokes Horace's "*dulce et decorum est pro patria mori*" (It is sweet and fitting to die for one's country), which is a purely nationalist sentiment. The poem "translates the personal hysteria of madness into the broader cultural hysteria of war fever," according to Markovits (156), but this translation shows the narrator embracing the "doom assigned" (III.59, 142), or in other words his status as self-sacrificing warrior going into battle for his country. He is, like the Light Brigade, charging into the valley of death. In embracing war, "Maud" follows the same trajectory as "Locksley Hall" when the narrator leaves to follow the bugle summoning his comrades and goes "roaring seaward" (l. 194, 63). The narrator's father in "Locksley Hall" died in "wild Mahratta-battle" (l. 155, 62) and he seems destined for the same fate. In both cases disappointment in love leads men to embrace death in war as recompense, finding solace in self-sacrifice in battle.

The label *The Valley of the Shadow of Death* for Fenton's photograph indicates that it had to be understood as a religious sacrifice in which the victims are killed to serve a higher purpose, and not as conventional military strategy. The images and accounts were framed both by national and religious ideologies that made the loss acceptable because the death of the soldiers fulfilled a higher destiny. The charge entered the popular imaginary as a heroic failure that served to reinforce a sense of national superiority, part of the British mythology of "heroic failure" analyzed by Stephanie Barczewski. Nationalism thus made the loss of life in the Crimean War acceptable and informed the elegiac tone of images such as Fenton's *Valley of the Shadow of Death*, even though no bodies were visible.

Photography in the American Civil War included dead bodies and operated under a different set of circumstances than those of Fenton's images of the Crimean War. Stereoscopic images created by dual-lens cameras were at the height of their popularity in the United States, instead of the single lens used by Fenton (Godbey 267). Also, most viewers were not introduced to the images at the *Dead of Antietam* exhibit at Mathew Brady's New York gallery in 1862 or Alexander Gardner's *Photographic Sketch Book of the War* (1866), because there were few galleries available to the public (Godbey 268), and the printed book was prohibitively expensive (267). Most people would have seen the images on postcards that were available cheaply through the mail (268), and they would have viewed them at home. Emily Godbey links the consumption of these photographs to the market for erotic imagery in trying to understand the simultaneous attraction to them and repulsion at the contents and the resulting "pleasurable fright from the safest of distances" (269). The reference to "at a distance" indicates that consuming these images at home made war a virtual experience that prefigured later media such as video from the war zone, and framed the experience in a way that insulated the viewer from the horror being depicted. Godbey uses the term "mediated" several times (270, 271, 273) to explain how the stereoscopic technology blunted the realism of the gruesome imagery in the viewer's awareness of the mediating device. The consumption of images of dead bodies in Civil War photography was therefore closer to the later media representations of war discussed in Chapter 5 than to Fenton's more anodyne photographs of the Crimean War.

Fenton's photographs reflect the frame of heroic masculinity around the representation of the Crimean War while it was being fought. After the cessation of hostilities, the effect of the dispatches by Russell and publications by Florence Nightingale helped shift the discourse around military masculinity to concern over the soldier's body and its care. The monument to the Crimean War erected in Waterloo Square in 1861 and the paintings of Elizabeth Thompson, Lady Butler registered the change in focus from the heroic to the wounded soldier's body. In the 1870s, images of the wounded soldier in sculpture and painting allowed the British public to come to terms with a war that ended with neither a definitive defeat nor complete victory after the occupying forces left Sevastopol and the Crimean Peninsula as part of the peace treaty. The wounded soldier's body became the vehicle for a recuperative discourse about the military failure of the Crimean campaign.

CHAPTER 2

The Soldier's Body and Sites of Mourning

After the end of the Crimean War in 1856, representations of the soldier's wounded body became the vehicle through which the British public came to terms with the ambiguous legacy of the conflict with Russia. The charge of the Light Brigade remained central to the legacy of the Crimean War, with commemorations of the action lasting into the late nineteenth century. Where images of an aggressive, self-sacrificing masculinity dominated in the first wave of public descriptions of the charge of the Light Brigade, the wounded soldier's body became an object of compassion in the memorials and paintings that followed the cessation of hostilities. Concern for the physical trauma of the soldier functioned as a balm for a wider national trauma. This shift in Victorian sentiment has parallels with that of the United States after the Vietnam War, when meaning was rescued from a conflict that did not result in an unambiguous victory, through the image of psychologically wounded veterans (see chapter 4). Veterans in both instances became surrogates for the process of national reconciliation after a setback where a country that saw itself as globally preeminent had to come to terms with failure.

The Crimean War for the British was a painful and ultimately inconclusive conflict that achieved little in its ambition to curb the southward expansion of Russia. The combined forces of Britain, France, and the Ottoman Empire succeeded eventually in capturing Sevastopol, but returned the city to Russia as part of the peace treaty that brought hostilities to an end. Queen Victoria said that "I own that the peace rather sticks in my throat" at the conclusion of the war, and this sentiment was underscored by the absence of victory parades and official welcoming ceremonies for the returning troops (Figes 467). Tennyson's poem on the charge is also a "narrative of loss" rather than of victory (Godfrey 4). Like the charge, the supposed "victory" of the Crimean War in the taking of Sevastopol was experienced as a loss, but through celebration of the charge of the Light

Brigade and other Crimean War incidents, the failure was turned into a symbol of British heroic failure. As Barczewski has argued, there is "a strain in British culture that embraces the nobility of suffering, defeat and heroism in the face of disaster over triumphalism and the glory of victory" (4). The image of wounded British soldiers reinforced the "nobility of suffering" that enabled the acceptance of the charge of the Light Brigade and the Crimean War as a whole as examples of heroism in defeat.

Along with the reporting of Russell on the horrific conditions suffered by British troops, the writings of Florence Nightingale helped focus attention on the body of the British soldier. Nightingale became a mythic figure during the Crimean War thanks to her organization of a nursing corps to tend to wounded soldiers. Nightingale's writings on the care of soldiers show an awareness of their bodies as neglected human capital, using language that recalls Michel Foucault's concept of "biopower": "That a man must be kept alive to fight, and cannot, if in hospital, be in the trenches, would seem to be a truth, though a truth not recognized, in general, by military authorities, though certainly Napoleon considered that the principal qualities to make a great military commander were civil ones" (McDonald 610).

Foucault describes biopower as a "new mechanism of power" in that it "exerts a positive influence on life, that endeavors to administer, optimize, and multiply it, subjecting it to precise controls and comprehensive regulations" (*History of Sexuality* 137). Nightingale exemplifies Foucault's concept of biopower because she was a hospital administrator who applied hygienic controls that enhanced the life expectancy of soldiers as a pool of bodies for the army. Care of the body extended the lives of soldiers, but also increased the firepower of the army by making them able to return to battle sooner. Nightingale's criticism of the army's neglect of the soldier's body in terms of food and clothing as well as the kind of care given to the wounded in hospital was scathing. While the mythologized figure of Nightingale shows her as "the lady with the lamp" in a hospital ward, she was concerned not only with the treatment of the wounded but also with the prevention of disease in soldiers by improved administration and regulation. Nightingale was primarily an administrator and reformer, and images showing her caring for soldiers in the hospital at Scutari place "a deceptive emphasis on Nightingale's work as personally healing soldiers" (Furneaux 205). Furthermore, she viewed the occupation of soldier not as a heroic calling but as employment on a par with any other job. She wrote that being a soldier was a "trade," and one that required a healthy body: "Yet upon these things [supplies of food and rations] depends the health of a man, and in the army, above all other trades, the physical state of the employed is the force of the employer. The health of the men is the main engine of the commander" (McDonald 612).

The economic language used by Nightingale shows how she perceived the soldier's body as a form of human capital, and in this case as an "engine" or source of power that drove the military as machinery. The soldier's body in Nightingale's terms becomes a resource to be cared for and ministered to for very practical and implicitly economic reasons. Nightingale's pragmatism leads her to view the military as an employer with certain basic responsibilities for employees, and the biocapital of the soldier's body as the commander's main natural resource. Nightingale's emphasis on the care of the body in her writings coincided with the use of the soldier's wounded body as a way of coming to terms with the ambiguous outcome of the Crimean War. This process was carried out symbolically through the bodies of the guardsmen on a war memorial in Waterloo Place, London, and the paintings of Lady Butler.

Memorializing the Dead

A monument to the Crimean War was erected in Waterloo Place in London 1861, and it reflected the prevailing attitude to the war at that time in that it "honours sacrifice rather than victory" (Berridge).[1] This was a controversial monument because commentators objected to the way in which it broke with established norms regarding war memorials. An article on "The Guards' Memorial in Waterloo Place" in the *Illustrated London News*, for example, expressed disdain for the memorial, calling it "an eyesore" and lamenting the "hideousness of the granite pile" (April 13, 1861, 330). The figure of Honour stands on top of the monument, bestowing garlands on the three guardsmen below (Figure 3). The commentator suggested that this figure was ambiguous, "its character and vocation being very problematical . . . sometimes suggesting the idea to the irreverent multitude of a street acrobat throwing his four rings" (330). The writer also found the figures of the guardsmen wanting, saying that if the architect John Bell had "thrown a little action into the latter figures, the absurdity and incongruity at present so conspicuous in the work might have been avoided" (330). The figure of Honour looks downward and downcast, as do the guards, suggesting mourning rather than victory; the overall effect was too somber for the reviewer for the *Illustrated London News*, who, it seems, would have preferred a more obviously victorious pose in keeping with the conventions of previous war memorials.

While the contributor to the *Illustrated London News* couched criticism of the monument in aesthetic terms, the objections to the monument have a subtext that depends upon the conventional depiction of masculine heroism and self-sacrifice analyzed in the previous chapter. The "action" that the commentator wished to see was related to ideas of men in battle proving their manhood and winning victory, but Bell's monument is actually an accurate depiction of British attitudes toward the Crimean War in that period. The soldiers in the Crimean memorial have their weapons at rest, with their heads bowed as if at a funeral rather than in the usual upright martial pose of other monuments. Near the Crimean memorial is a monument to Prince Frederick, Duke of York (1834),

Figure 3. The Waterloo Place monument from the *Illustrated London News*. Courtesy of Brock University Special Collections.

which was made to be seen by pedestrians when descending from Waterloo Place to Carlton House Square (Yarrington 246–276). The memorial places the Duke of York atop a 125-foot (38 meters) column, the height expressing his class status and "loftiness" (Abousnnouga and Machin 42); the statue also shows him in a forceful pose, with a sword at his side as if surveying the scene of battle. The angle of his head and imperious gaze suggest command, in contrast to the mournful poses of the Crimean monument.

The bodies represented on the Crimean War monument are therefore markedly different from the heroic pose of such statues as that of the Duke of York. The monument was financed by commission by the Guards themselves, which accounts partly for the emphasis on their unit, but it also caught the spirit of the belated commemoration of the Crimean War. It was erected in 1861 and was a little ahead of its time; Paul Usherwood and Jenny Spencer-Smith note that it was "not until 1875 that the first major Balaclava reunion for veterans of the Light and Heavy Brigade charges was held at Alexandra Palace and two years later the Balaclava Commemoration Society was formed" (34), so that for two decades the Crimean campaign was a neglected war. Indeed, when Lady Butler announced her decision to paint a scene from the Crimean campaign, her father asked why she wanted to paint that "forgotten war"? (Elizabeth Butler, *Autobiography* 101). The emphasis on the ordinary soldier as an iconic figure, however, prefigured the recuperation of the Crimean War through icons such as the guardsman, especially in Butler's art.

The bodies of guardsmen figured prominently in a painting based on the Crimean War that was exhibited in 1874; the enthusiastic reception of *Calling the Roll after an Engagement in Crimea* (commonly referred to as *The Roll Call*) catapulted Butler to fame. While preparing for the painting, she read Russell's dispatches (Usherwood and Spencer-Smith 67). In contrast to Fenton's photographs, Russell's reporting described the physical damage of combat, as well as the toll taken on soldiers by their inadequate shelter and lack of food. Russell's reporting also conveyed the aftermath of battle, with wounded and dead soldiers scattered across the landscape:

> One could not walk for the bodies. The most frightful mutilations the human body can suffer—the groans of the wounded—the packs, helmets, arms, clothes, scattered over the ground—all formed a scene that one could never forget. There, writhing in their gore—racked with the agony of every imaginable wound—famishing with thirst—chilled with the night air—the combatants lay indiscriminately, no attempt being made to relieve their suffering until the next day. (53)

Russell's emphasis on soldiers' bodies is echoed in Lady Butler's painting. Like Russell's reporting, Lady Butler's painting depicts the aftermath of war through damage to bodies, with one soldier prone on the ground and others visibly wounded and fatigued.[2] The painting is remarkable for the way in which it substitutes for the heroic upright masculinity images of wounded and prone male bodies. The enthusiastic reception of the painting indicates that it appeared at a time when the British public was receptive to an image of soldiers as vulnerable and in need of care. *The Roll Call* was praised for its "realism," but its acceptance suggests that visually, like Fenton's *Valley of Death* photograph, it implicitly values sacrifice even as it represents death and loss in war. Countervailing signs that justified the soldiers' sacrifice were incorporated in the tableau, including upright standards, fleeing Russians in the distance, and a blood-spattered Russian helmet in the snow, all of which suggests that their sacrifice was not in vain. The painting has a subtext of the nobility of sacrifice, because Lady Butler was impressed by the stoicism of the British soldier and esteemed praise for her painting from Crimean veterans the highest honor (Usherwood and Spencer-Smith 67). As Usherwood and Spencer-Smith say, after Lady Butler's painting, "the concept of the manly soldier, valiantly bearing the marks of recent conflict, became an essential feature of the genre" (59); their formulation "the marks of recent conflict" underscores how the wounded soldier's body replaced the heroic warrior, and their use of "manly" how the guardsmen in the monument were representative of British military masculinity. Manliness is represented by a wounded soldier's body rather than the figure of a charging cavalry officer.

Lady Butler's approach to painting paralleled the attention to the soldier's body in other areas. She described her method as painting the body of the soldier first, and then the greatcoats and other clothing on top afterward (Elizabeth Butler, *Autobiography* 102), in a visual analogue to Florence Nightingale's focus on the soldier's body. Lady Butler wrote that "war calls forth the noblest and basest impulses of human nature" (47), and her representation of the wounded body combines these antithetical values. Lady Butler admired the fortitude of the soldiers she painted, but also acknowledged the damage to bodies from the organized violence of combat. Rather than the "heroic failure" of the accounts and poems on the charge of the Light Brigade, her paintings show "heroic loss" and allow for the mourning that was curtailed by the ambiguous ending of the Crimean War. The reviews of Lady Butler's painting paralleled those of Luke Fildes's *Applicants for Admission to a Casual Ward* (1874), reflecting a "social reformist climate" receptive to appeals to sentiment (Usherwood and Spencer-Smith 33). As Furneaux says, "reforms to accommodation, discipline, recreation, health, and education" were enacted after the Crimean War as the military itself accepted a "caring paternalist ideal" (133). Anne Summers argues that in this environment "the sick and wounded private soldiers, portrayed in the most favourable light, were glorified as the finest types of their social class. In a sense they were treated as a special category of 'the deserving poor'" (105). The figures of the guardsmen on the Waterloo Place monument elicit commiseration thanks to their downcast pose, paralleling the portrayal of soldiers in general as deserving of sympathy in the same way as the poor. Military masculinity was thus redefined in the collective mourning after the Crimean War, reflecting awareness of the soldier's body as vulnerable, as introduced in Nightingale's texts. In Judith Butler's terms, it became possible to acknowledge soldiers' precarity, which put them on a par with other vulnerable populations.

This sentiment is captured in a *Times* review of *The Roll Call* that called it "a picture of the battlefield, neither ridiculous, nor offensive, nor improbable, nor exaggerated, in which there is neither swagger, nor sentimentalism, but plain, manly, pathetic, heroic truth" (quoted in Usherwood and Spencer-Smith 31), underscoring how it was interpreted as reaffirming masculinity even as it showed a prone body. Masculinity was represented through the soldier's wounded body in this instance, but in another of Lady Butler's paintings the central figure controversially exhibited the mental rather than the physical effects of war trauma. While war was seen in Victorian images as wounding the body, it was not yet usually represented as mental trauma, as it is in the contemporary diagnosis of post-traumatic stress disorder (PTSD) (see chapter 4).

The Charge of the Light Brigade and Psychological Trauma

While *The Roll Call* focused on the wounded body, Lady Butler's painting commemorating the charge of the Light Brigade controversially also showed a soldier traumatized by battle. In *Balaclava* (1876), Lady Butler tackled the aftermath of the charge of the Light Brigade rather than trying to re-create the charge itself. The painting is remarkable for the way in which it represents the wounds of the cavalry and also attempts to show both the physical and mental trauma of combat. The central figure faces the viewer directly, looking dazed. The veteran portraying the central figure, Private William Henry Pennington, received much criticism for his portrayal of a traumatized soldier. Pennington was accused by the *Manchester Critic* of being "theatrical—not dramatic—simply ruinously obtrusive and unreal," and by another critic of being "dazed and drunk with wine of battle" (quoted in Usherwood and Spencer-Smith 65). Both reactions, in their different ways, show discomfort with Butler's use of Pennington to show the psychological aftereffects of battle, and her divergence from the stoic expressions of *The Roll Call*. Calling him drunk suggested he was not in his right mind, although "wine" and inebriation are misleading euphemisms for the effect of violence on the soldier. Rather than acknowledge the possible mental aftermath of warfare, the comments trivialize the effects as the result of drunkenness, with the implication that the soldier was at a social occasion and over imbibed rather than returning from the battlefield.

The criticism that Pennington was being "dramatic" may have been based on the knowledge that he was an actor. While Pennington was indeed an actor when

Figure 4. Lady Butler, *Balaclava*. Courtesy of Manchester Art Gallery/Bridgeman Images.

he posed for Butler's painting, he was also a survivor of the charge of the Light Brigade and recounted his experience in two autobiographical texts, *Sea, Camp, and Stage: Incidents in the Life of a Survivor of the Light Brigade* (1906) and *Left of the Six Hundred* (1887). As both titles indicate, Pennington constructed an identity and a career around the image of himself as a survivor of the charge of the Light Brigade. After his discharge from the army, Pennington dabbled in what he termed "scholastic work" (*Sea* 25), but realized that his true vocation lay on the stage, and the bulk of his memoir describes his various theatrical performances in detail. His association with Lady Butler started at the first commemoration of survivors of the charge at the Alexandra Palace Monday, October 25, 1875. The celebration became a nexus of images of the charge, with memorabilia on display and Lady Butler's most famous painting lent by the queen for the occasion:

> Miss Elizabeth Thompson (now Lady Butler, the renowned painter of scenes from military life), to whom I was about to give sittings for the central figure of her "Balaclava," entered thoroughly into the spirit of the promoters of the fete, and her pathetic picture, "The Roll Call," . . . was not among the least of the attractions offered to the admiring gaze of the throngs gathered from far and near. (*Sea* 145)

Pennington participated by giving a rendition of Tennyson's "The Charge of the Light Brigade" (although he was irritated that he was not given top billing in advance notices of the event), which became his signature performance. By 1875 Tennyson's poem had become the lens through which to understand the

Crimean War as a whole, and the fact that the celebration in 1875 focused solely on the survivors of the charge rather than veterans of the entire conflict demonstrated its central importance in the public imaginary. Lady Butler built on the centrality of the charge in her painting *Balaclava*, which, although it takes the battle as its title, is concerned solely with the Light Brigade. Pennington takes center stage in the painting as the preeminent dramatic performer of the charge.

Pennington incurred Lady Butler's wrath when asked subsequently to recreate the role in a *tableau vivant* for her; he did not wear his busby in the right attitude and ruined the effect of the piece, because in Lady Butler's mind he gave a performance of "Mr. So-and-so in the becoming uniform of a hussar" rather than "my battered trooper" (155). Lady Butler in her art was aiming at the illusion of reality, and she felt that Pennington was too obviously calling attention to himself and his pride in his uniform in the tableau vivant. Whereas in that case Pennington was not sufficiently a "battered trooper," in the painting Pennington, with his dazed expression and sword hanging at his side, was too obviously portraying mental anguish. By trying to create the mental as well as the physical injuries of battle, Pennington strayed beyond the focus on the soldier's body by drawing attention to the effects of combat trauma on the mind.

Pennington's dramatization of mental trauma also transgressed the image of what Jill Matus terms the "determined stoicism" of the British soldier (49). The ideology of heroism made displays of mental trauma a transgression of what one memoirist termed "that soldierly pride which forbids outward signs of disclosing self-pity or despair" (Matus 49–50).[3] Pennington's "battered trooper" violates the norms of stoicism expected of British troops. Matus does not comment on the gendered implications of such passages, but these strictures apply only to men. The prohibition against expressions of emotion ensured silent obedience to blunders such as the order for the Light Brigade to charge directly at the Russian artillery. Pennington's performance for Lady Butler's painting exposed the unheroic but common reaction of a soldier to the trauma of battle. The negative reaction of the commentators, who in general praised the painting's overall effect, show that Pennington violated the norms of heroic masculinity in representing a mental rather than a physical wound.

The censure of Pennington indicates that, while there were examples of "military men of feeling," as Furneaux has argued, these emotions were expressed privately, and that the tolerance for emotionality in men was circumscribed in public. Letters from soldiers in the Crimean campaign registered the horror and suffering of war and expressed grief, but these were private and written for intimates. The situation was contradictory, because while "the military figure most applauded was notable, not for a stiff upper lip, but for a particular capacity for feeling" (Furneaux, "Victorian Masculinities" 218), the memoirs cited by Matus by contrast show that war was viewed as a "stimulant" (117), with an emphasis on masculine actions and bravery rather than expression of emotions. Men undoubtedly did experience emotions in war, but they did not find their way into popular representation of military masculinity. Also, while Pennington dramatized the traumatic effects of combat in performance, his memoirs give a straightforward account of his participation in the action and his good fortune in emerging lightly wounded from the "mad, desperate charge" (*Left* 3). Indeed, he credits his wound with sparing him the horrors of the winter (*Left* 61), and on his return to England he records no lingering mental effects from his participation in the charge. Pennington the author was not as troubled by trauma as his portrayal of a survivor of the charge would suggest.

Ironically, Pennington himself was critical of a theatrical attempt to re-create the charge when he performed in a play written by a friend that was designed as a vehicle for his signature performance of Tennyson's poem:

> I believe it was in the autumn of 1878 that a drama entitled "Balaclava," which had been somewhat hastily contrived by my old colleague J. B. Johnstone, was most inadequately presented at the Standard Theatre upon the off-chance of a few successful nights, but not a penny was expended on the piece. The title was possibly expected to achieve everything. (*Sea* 166)

Pennington felt that the author relied too much on the name Balaclava and the mention of the charge of the Light Brigade in advertising to entice an audience, and did not spend enough time or money on staging the battle. He was especially caustic about the playwright's attempt to convey the charge itself:

> There was no attempt to realise the "Charge," and in this impossibility, of course, lay much of the weakness of the play. The advance of the "Six Hundred" was described by George Byrne (who appeared as a military correspondent posted upon an elevation near the R.U.E.) in spasmodic shouts in the intervals between the heavy discharges of artillery and musketry, which threw the people in the vicinity of the theatre into a fever of alarm. (*Sea* 166–167)

Pennington felt that there should have been more of an effort to reenact the charge on stage, although he does not specify how exactly the production was supposed to achieve this effect with a small cast of ten actors trying to reproduce a cavalry action against Russian artillery. Pennington also criticized the play for a mistake in the uniforms, saying that the only one "at all near the mark was my own" (167), which he apparently found embarrassing since several survivors of the charge were in the audience. He also felt there were not enough extras involved, because the "numbers upon the stage doing duty for the heroic 'Six Hundred' were by no means worthy representatives of that historic band" (167); presumably, Pennington did not want six hundred actors and horses on stage, but he does not indicate how many more representatives of the six hundred would have satisfied him. Overall, his comments show how he viewed himself not only as a veteran of the charge but a guardian of the reputation of survivors and of the verisimilitude of reenactments of the charge as a "historic band," with a special status that deserved exceptional treatment when depicting their exploits.

Feelings about the Crimean War were channeled through reenactments of the charge like the play *Balaclava*, especially in the focus on individual participants as representative of the war as a whole. Fenton's photographs showed individual soldiers, and both of Lady Butler's paintings, in contrast to other renderings of the charge such as Barker's, focused on a central figure to convey the pathos of the aftermath; the play *Balaclava* turns the aftermath into a psychological drama. While Pennington criticized J. B. Johnstone's play, it paralleled his own performance in Butler's *Balaclava* in the representation of the mental damage of war. Frank Walton, the hero of the story, joins the cavalry both to pay off a debt and to win back his sweetheart, but when he returns to England he is traumatized after having participated in the charge of the Light Brigade. He has amnesia and is only able to recite lines from Tennyson's "Charge of the Light Brigade" in response to questions from his family. In keeping with Pennington's portrayal of the traumatized survivor for Butler's painting, the character he plays cannot escape the memory of the charge as the defining moment of the Crimean campaign.

The play's opening indicates Frank's future status as soldier when he is described by Tabitha, the domestic servant, as having "fought the battle nobly" (4) in trying to save the family farm from the bankruptcy it now faces. Frank resolves to "be a man" (8) and do what it takes to restore the family fortune, combining the tropes of masculinity and warfare; it is inevitable, therefore, that he eventually decides to enlist in the army for a year and to leave for a place ". . . where men play the desperate game of life and honour, where there is much to win and much to dare, for where the proudest have gathered glory and a name, where a bold heart wins a future lighted by a country's gratitude and a country's thanks" (21).

Words such as "honour" and "glory" make it clear that Frank is leaving for the Crimean War, which, although it is never named as such in the play, is indicated in the title. Key words such as "honour" and "glory" were indelibly linked to the Crimean War, and especially the charge of the Light Brigade, by Tennyson's poem. It is also assumed that to be "manly," a British male must inevitably fight in a war. Frank affirms his masculinity by enlisting in the war and fighting for the hand of Annie Blair, the local milliner's daughter. Squire Harold, the dastardly villain of the piece, has made Annie's hand in marriage a precondition for forgiving the debts accrued by Frank's father, but if Annie is still unmarried by the time he returns from war in one year's time, then she will have the option to marry Frank. Act I culminates with a patriotic speech from Frank as he prepares to leave for the Crimean War, which echoes the characterization of the charge of the Light Brigade as both self-sacrifice and madness in Tennyson's poem, "True, mother, it is the madness that sends forth the power and strength of England to contend against the despotic and barbarous tyranny that would oppress and destroy. It is the madness that bands men in the holy cause to curb and check ambitious selfish sway. It is the madness that protects the weak and defies the strong" (29).

The speech uses the conventional tropes of English nationalism, with the country as the bastion of liberty opposing the forces of oppression and despotism. The appeal to "madness" also recalls the characterization of the charge of the Light Brigade as a kind of insanity (see chapter 1), but the play, like Tennyson's poem, turns this into an admirable quality that is enlisted in a heroic act. The emphasis on madness prefigures the effect of the charge on Frank when he returns suffering from a form of heroic insanity, both noble for having participated in the charge and suffering from mental derangement at the same time. Pennington's acting in the play echoes his portrayal of the shocked and disoriented charge survivor in Butler's painting, and the focus of the drama becomes the possibility of his recovery from trauma.

No portrayal of the Crimean War would be complete without the invocation of Florence Nightingale, so the plot has Tabitha Trust, the family's former domestic servant, reappear in Act II as a nurse. Tabitha is "another angel in petticoats," bending over the bed of a wounded soldier and "with the loving eyes of a sister" evoking images of the nurses who worked with Nightingale at Scutari (25). Tabitha has been inspired by Nightingale's example (26), but does not name her directly, referring rather to the sacrifices made by a great lady. The play also briefly references the initial suffering of British troops in the Crimean campaign, with complaints about the lack of food (24) and the inadequate tents provided by the army (26).

The real interest in Act II is the charge itself. To represent the charge, J. B. Johnstone has Munro, a sympathetic auctioneer in Act I who helped save the family Bible for Mrs. Walton, reappear in Crimea as a "foreign correspondent" for his own newspaper that his wife is running for him in his absence (31). Munro is a surrogate for Russell in that he describes for the audience a charge that they cannot witness directly, with the stage directions calling for offstage noises to suggest a battle: "Why they are never mad enough to send that handful of men against the myriads that line the hill side and cover every vantage ground that lie between them and the guns! (loud huzzas as at a charge sounded, distant guns and reports of muskets)" (32).

Munro's narration echoes the "wild" charge from both Tennyson's poem and Frank's rousing speech at the end of Act I. Munro also reinforces the idea that the charge of the Light Brigade is a suicide mission by exclaiming that he is going to have to report news about Frank's "courage and your death" (32). However, Frank staggers back from the charge wounded, and proceeds to give his own account of the action, saying that "the enemy's guns were before us and it was our duty to take them," and "on we rushed with sabres flashing," using the vocabulary of "duty" and the conventional metonym of cavalry as flashing weapons found throughout poetry describing the charge of the Light Brigade (see chapter 1). The stage directions then call for Frank to "madly" relive the charge uttering "Ride 'em down. . . . Ha! Ha! Ha! Saved! The heavy brigade have covered our retreat" (34), prefiguring his reappearance in Act III as a traumatized survivor who is trapped in the past.

Act III draws out the tension by having Tabitha (now back from the Crimean War) and Annie discuss a newspaper account of the charge of the "six hundred" and the probability that Frank is dead. They reenact the experience of the civilian readership, who could only experience the charge vicariously through the dispatches of war correspondents like Russell. Eventually it becomes clear that Frank is not dead, although Munro, also returned from the Crimean War and vowing never to return there, won't comment on Frank's condition. Finally, Frank appears on stage "with a vacant look," to a scream from his mother and swooning from Annie; Munro then explains to them that the only event that Frank can remember from the past is the charge of the Light Brigade, and the hope is that bringing him home might restore his memory (47). From this point onward, Frank's response to any question posed to him is to recite lines from Tennyson's poem, allowing Pennington to deliver his signature recitation of the poem. While this may seem comic to a contemporary reader, it shows the power of Tennyson's poem that it could be assumed that Frank reciting the lines would elicit an emotional response.

Johnstone makes one critical error in his transcription of Tennyson's poem in writing that the Light Brigade had "to do or die" rather than "do and die," unintentionally undermining the portrayal of the charge in Tennyson's poem as a suicide mission (48). Tennyson suggests that the cavalry in obeying orders will inevitably die, whereas Johnstone's version suggests a possible alternative outcome. He also divides the poem into different sections to draw out the final scene, as it appears that the evil plan of Squire Harold to marry Annie may succeed because of Frank's mental incapacity; Squire Harold says in an aside to the audience, "his coming back isn't likely to cross my purpose" (48), which presumably was designed to elicit boos and hisses. While reciting the lines, Frank keeps putting his hand to his breast as if searching for something (which the audience knows is the packet that contains proof that Squire Harold defrauded his father), but it is not until much later in the scene that Munro reveals that he was with Frank in the hospital in Scutari and was entrusted with the evidence of the fraud, which he then reveals.

This revelation has the immediate effect of restoring Frank's memory and leads to Squire Harold confessing his crime. Within the space of a few minutes the script then rushes to a conclusion: after Squire Harold confesses and Frank says that he forgives him, there is a touching scene in which he and Annie agree that they will marry. Previously, Tabitha had returned to Mrs. Walton and said she could no longer be a domestic servant after her experience as a nurse in the Crimean War, but would be her friend (38), so that by the end of the play reconciliation and friendship mark every relationship. Frank then gives the final speech of the play, "Heaven has been with me in this, as it was in its mercy in the October fight when England's honour was maintained by English hearts, for the living we can but say as men they did their duty, for the dead reverence and respect. (removes hat. Recites) "when can their glory fade . . . noble six hundred" (54).

The recitation is designed to provide an emotional climax that fuses the scenes of reconciliation and forgiveness that preceded it with the memory of the charge of the Light Brigade, as memorialized in Tennyson's poem. Once again, the term "duty" binds the audience with the Light Brigade in an imaginative parallel to their own conduct (see chapter 1), as well as the usual appeal to nationalism and English exceptionalism. The play, like Tennyson's and other poems, builds an imagined community around the ideal of self-sacrifice for the nation as duty and self-subordination.[4] The overt emotional appeals in the play suggest that the same kind of emotional identification occurred when viewing Lady Butler's painting of what was left of "the noble six hundred," for which Pennington was a model, and his presence on stage at the end of the play reciting these lines underscores his status as both an actor and a living embodiment of the charge itself.

Pennington's role shows how the charge came to stand for the Crimean War as a whole and reconcile the Victorian public to its outcome as a heroic defeat. The charge was a military disaster, but Johnstone's play emphasizes eulogy and reverence, and the multiple reconciliations encourage a feeling of closure. The play also appeals directly to the middle classes by having Frank's recuperation of the family fortunes derive directly from his participation in the charge of the Light Brigade. The stage directions describe the setting as "a room in a middle class farm house . . . devoid of furniture, except two wooden chairs," showing the decline of the Walton household from middle-class status. By Act III, the same room is "moderately well furnished," presumably due to Frank's efforts, and his marriage to Annie will continue the Walton family line. The danger of falling down the social scale is averted at the end of the play as the family fortunes are restored, relieving a key middle-class anxiety in its conclusion. The play is thus reassuring on a number of levels, showing that manly valor can help restore the family fortunes and by extension national pride, even though the charge itself was a military blunder. However, Pennington's performance in the play and his posture in Lady Butler's painting radically reorient the representation of the wounds of war from the soldier's body to lingering mental trauma.

Diagnosing Trauma

The commemoration of the Crimean War and Pennington's role highlight the similarities and differences between the frame around Fenton's photographs,

Lady Butler's paintings, and the later photographs by Ristelhueber (see chapter 5). Pennington's case is noteworthy because trauma was conceived largely in terms of the body in the Victorian era. As Matus says in her excellent analysis of nineteenth-century concepts of trauma, for the word to encompass mental damage, "the mind had to be conceived of as physical, material and physiological—and therefore vulnerable—like the body" (7). The major terms for trauma at this time were "brain fever," "railway shock," and "railway spine," and they were coined at a time when the understanding of trauma was gradually shifting from the purely physical to encompass the psychological effects. The shift was incremental, and it was not until the twentieth century that "trauma" could be viewed as a purely mental phenomenon.

Terms like "railway shock" were produced by the collision of human bodies with increased mechanization, and prefigured later terms such as "shell shock" that tried to account for the psychological effects of the industrialization of war on soldiers' minds in World War I; indeed, the soldier has been characterized as "an industrial worker of sorts" (quoted in Luckhurst 59). Matus and Roger Luckhurst trace the gradual redefinition of injury in war from physical wounds exclusively to the recognition of mental trauma, especially the role played by artillery bombardments in diagnoses such as "shell shock."[5] This was far from a uniform process, as Tracey Loughran records (80). Loughran notes that there were two forms of "shell shock" recognized during World War I, "shell-shock W" and "shell-shock S," depending on whether soldiers were labeled as "wounded" or "sick" (11). Only those wounded by enemy action were entitled to "wound stripes and military pensions" (11), showing the continued bias toward trauma as bodily injury and the association of mental issues with cowardice. It was not until the creation of the category of post-traumatic stress disorder (PTSD) that the transition from war as bodily injury to a definition of war as a wound of the mind was completed in a final recognition of the syndrome. However, as I argue in chapter 4, this redefinition of war as mental trauma was framed by the suppression of wounded bodies in the media, and thus became another instantiation of the late twentieth-century representational regime of a "war without bodies." The late twentieth century also ushered in the paradoxical situation analyzed by Simons and Lucaites in *In/Visible War* that war was both ubiquitous and unremarkable. This situation was facilitated by the widespread adoption of war gaming, first in tabletop gaming and later in military-style video games, as I argue in the next chapter.

CHAPTER 3

War Games

Games, war, and imperialism were intertwined in the late nineteenth and early twentieth centuries. In *Postmodern Imperialism* (2011), Eric Walberg has termed this period when Britain was in the ascendant as the "Great Game I," because "the entire world was now a gigantic playing field for the major industrial powers, and Eurasia was the center of this playing field" (17). The term "great game" was coined by Arthur Connolly, a British intelligence officer, in a letter in July 1840 to describe possible relations between Britain, Russia, and Persia. Connolly meant "game" in a diplomatic sense, as enabling peaceful negotiations between the nations (Kaye, vol. 2 101). By the time Rudyard Kipling referred to the "great game" in *Kim* (1901) to describe the conflict between Britain and Russia over Afghanistan, it had much more imperial overtones and was used as a way to characterize Britain's jockeying for influence across the globe. Cecil Rhodes was the chief advocate of imperialism and is reputed to have said that "I saw that expansion was everything, and that the world's surface being limited, the great object of present humanity should be to take as much of the world as it possibly could" (Maguire 11); by this he meant that as much of the globe as possible was to be placed under British rule, which was his version of the "great game." As Charlotte Eubanks argues, "ludic (playful, game-based) fantasy often develops in tandem with geopolitical events" (36).[1]

A confluence of games and geopolitics can be found in H. G. Wells's books *Floor Games* (1911) and *Little Wars* (1913), in which he codified the rules for playing at war while mirroring the idea of the world as a playing field available for British expansion. *Little Wars* in particular created the paradigm for later imaginary battle play such as miniature war gaming, Dungeons and Dragons, and ultimately military strategy video games. Tracing this lineage shows the increasing confluence of fantasy and war play as the lines between military uses of strategy games and civilian imaginary participation in war expanded. Military strategy games in particular, with their emphasis on conquering and controlling territory, further a belligerent and pervasive expansionism that, while it is not as overtly racist and hegemonic as Rhodes's pronouncements, in their assumptions about the availability of territory to conquer still rely on an imperialist worldview.

While Wells himself overtly opposed imperialism, in *Floor Games* he created a miniature version of the "great game."[2] The topography of the games included a "civilized" center and an underdeveloped periphery that followed the Rhodes model of Africa as a resource to be exploited by the British. Wells and his sons built imaginary civilizations that, not surprisingly, reproduced Victorian society writ small (10–12). The civilized islands were populated by a bureaucratic government of mayors in town halls, train stations with porters, a "motor omnibus," an inn, and a zoo. These islands represented Victorian "civilization," whereas the other two islands were undeveloped and apparently available for exploitation. One island contained "great mineral wealth" (7), as well as being of "exceptional interest to the geologist and scientific explorer" (8). Wells makes an ironic comment about one of his sons having "Imperialist intentions" (8), but the fantasy created here does in fact align with the expansionist ethos of the British Empire. The scenario is that more developed countries are able to sail to and exploit the resources of less developed countries, which in Wells's fantasy landscape are inhabited by "negroid savages" (6) and by "Indians" living in tents (7). Wells and his sons had a dominant navy and were free to roam the "oceans" between their imaginary kingdoms in a replication of British global naval superiority. There was an exception for Zulus, for whom there were "special rules" (3), although it is not specified what these are, but presumably this is a reference to the Anglo-Zulu War of 1879 and the cultural memory of the British army's defeat in the Battle of Isandlwana, which established the Zulus' reputation as fearsome warriors.

The presence of a zoo also marks British dominion over animals (13). Wild animals were collected and exhibited at the Regent's Park Zoo, which was envisioned as "a collection of captive wild animals that would serve not just as a popular symbol of human domination, but also as a more precise and elaborate figuration of England's imperial enterprise" (Ritvo 206). Sarah Amato too sees the zoo as part of the imperial "civilizing" process whereby the superiority of British culture was implicitly reinforced (112). The Zoological Society and the zoo were thus directly linked to imperialism and the domination of other countries through animal bodies.[3] The Wells fantasy zoo contained a motley assortment of animals from across the globe, including camels, elephants, pigs, parrots, and bears. The range of animals in the game was constrained by what was commercially available for purchase, and tended toward exotic rather than native British species. As in Regent's Park Zoo, Wells included animals representing the reach of the empire, showcasing the intersection between an imaginary topography and the wider cultural intersection of imperialism and collecting of specimens. The islands in Wells's *Floor Games* thus represented the "cultural ephemera of imperialism" (quoted in Brown 237) so that, while Wells may have been critical of the British Empire in his fiction, such as *The War of the Worlds* (1897), his games drew on popular imagery that implicitly endorsed British expansionism.

Wells lamented not being able to find noncombatants for his game, writing that "we want civilians very badly" (*Floor Games* 3). There were beefeaters and an abundance of railway porters to go with a train set, as well as a German set where "even the grocer wears epaulettes" (3), but otherwise all the figures were soldiers. Wells wished he could buy tradesmen and servants, because "with such boxes of civilians we could have much more fun than with the running, marching, swashbuckling soldiery that pervades us" (3). His lament over the lack of civilians shows that his first imaginary games encompassed other activities beyond the military, but that only soldiers and figures that represented activities such as transportation and commerce were available for purchase.

With his next book, *Little Wars*, the lack of civilians ceased to be a problem. The focus in *Little Wars* is completely on "swashbuckling soldiery," because Wells and a friend created the rules for mock battles using toy soldiers exclusively. *Little Wars* is credited as one of the most influential war gaming books aimed at the general public. There were guides to military gaming that preceded *Little Wars*, but they addressed the rules for *Kriegsspiel* and were of interest mainly to the military, as Christopher Yi-Han Choy has documented in "British War-Gaming, 1870–1914."[4] Wells's text was published amid a boom in the production of toy soldiers following an advance in technology by the William Britain Company. Where previously toy soldiers had been made of solid metal, the company started manufacturing hollow metal figures, thus making their product cheaper to manufacture and ship than their rivals (Brown 238). Wells in *Little Wars* extols the virtues of war gaming over actual violence:

> And if I might for a moment trumpet! How much better is this amiable miniature than the Real Thing! . . . Here is the premeditation, the thrill, the strain of accumulating victory or disaster—and no smashed nor sanguinary bodies, no shattered fine buildings nor devastated country sides, no petty cruelties, none of that awful universal boredom and embitterment, that tiresome delay or stoppage or embarrassment of every gracious bold, sweet, and charming thing, that we who are old enough to remember a real modern war know to be the reality of belligerence. (52)

This is Wells's version of "war without bodies," where the toy soldiers replace the real bodies of soldiers. There are several problems with Wells's sanguine statement about the superiority of imaginary warfare, however. As Kenneth Brown has argued, the proliferation of toy soldiers like those that Wells deployed helped create a culture in which war permeated the civilian sphere and formed "part of a complex web of educative influences, both formal and informal, which linked games on the nursery floor to the adolescent and adult worlds" (243). The toy manufacturer Britains Ltd. came out with a catalogue in 1908 entitled *The Great War Game for Young and Old*, which juxtaposed their toys with real-life military equivalents, suggesting that games and war were intertwined (Brown 247). While Wells might hope that his war games were distant from the real thing, other forces in the culture insisted on their proximity, undermining his pacifist intentions. Where Wells emphasized their difference, the manufacturer played on the proximity of imaginary to real warfare, thus reinforcing the connection between the two.

As Nigel Lepianka and Deanna Stover have argued, Wells himself was ambivalent about war games. He admitted in his *Experiment in Autobiography* (1932–1934) that his early life was marked by "war fantasies," but that he had lost his enjoyment of imaginary warfare thanks to World War I, writing that "up to 1914, I found a lively interest in playing a war game, with toy soldiers and guns, that recalled a peculiar quality and pleasure of those early reveries" (74). For Wells, as for many other people, the experience of World War I made it hard to romanticize organized violence in the wake of industrialized warfare that increased military firepower and led to mass casualties. In their emphasis on close combat and fair play, the games described in *Little Wars* were nostalgic for premechanized warfare, but Wells himself had predicted the obsolescence of such tactics brought about by advances in military technology, such as tanks in *The Land Ironclad* (1903), aerial warfare in *The War in the Air* (1908), and atomic bombs in *The World Set Free* (1914) (Lepianka and Stover). While warfare was becoming increasingly mechanized and enabled killing at a distance, *Little Wars* simulated conditions more like those under which the Light Brigade operated in the Crimean War, including cavalry charges (43), showing his continued allegiance to a romanticized version of battle rather than combat transformed by technology.

Wells was not the only one drawn to war games in this era; Robert Louis Stevenson created his own "war game in miniature" (Brown 245), and Wells played his game against a number of prominent writers and politicians (241). The interest in war games was also an index of the increasing presence of militarism, not only in British society but in Europe generally. "Militarism" was initially used in the 1860s as a pejorative term applied to Prussia and France, which were seen as excessively geared toward expanding their armed forces, and it was used as the antithesis of cherished British liberty, as it was in the poetry on the charge of the Light Brigade (Stargardt 13), but the acceptance of war gaming indicates a shift in attitudes.[5] The military and noncombatants were linked previously when civilians would accompany soldiers on campaign, as Becky Sharp does in William Thackeray's *Vanity Fair*, for instance, but war became much more part of the domestic civilian imaginary thanks not only to toy soldiers but also to faster reporting from the battlefield by figures like Russell (see chapter 1) and to the expansion of *Kriegsspiel* beyond military circles to war gaming as a hobby. Despite the earlier criticism of militarism in Britain in the nineteenth century, by the early years of the twentieth century the country was itself in the midst of an arms race with Germany, especially in naval power. Writers like Thomas Hardy saw militarism as the chief cause of World War I, and blamed "glorification of war and the warrior" for the violence (Wickens 415–424); the increased production of toy soldiers was a marker of this commercialization of warfare for entertainment, prefiguring the effect of military-style video games in the later twentieth century.

War was never far from Wells's mind, even if only as fantasy. When writing *The War of the Worlds*, Wells cycled around the town of Woking where he lived and imagined aliens destroying it, as he says in the "Introduction" to the text, "The scene is laid mainly in Surrey in the country round about Woking, where the writer was living when the book was written. He would take his bicycle of an afternoon and note the houses and cottages and typical inhabitants and passers-by, to be destroyed after tea by Heat-Ray or smothered in red-weed" (ix–x). *The War of the Worlds*, like many of his texts, imagines the effect of superior military technology like the Heat Ray. As Peter J. Beck says, "the narrator covers fast-moving events rather like a war correspondent submitting regular reports" (158), underlining the proximity of the fantasy battles in *The War of the Worlds* to the reports from the Crimean War by Russell. For Wells, fantasy warfare that mixed violence and the ludic were connected with games in his version of "warplay" (Filewod 17), anticipating the combination of elements in tabletop and video war games. The ludic was part of writing a fantasy conflict like *The War of the Worlds*, which he describes in terms of a game, "Nothing remains interesting where anything may happen. For the writer of fantastic stories to help the reader to play the game properly, he must help him in every unobtrusive way to *domesticate* the impossible hypothesis" (*Seven Famous Novels* vii–viii; italics in original). The mundane and the fantastic in his novels are sutured by "playing the game," which, as in *The War of the Worlds*, mixes realistic settings with imaginary violence.[6] Wells's use of the metaphor of game recalls the "great game" of imperial geopolitics and links it to his own fantasies of war. The same procedure is at work in *Little Wars*, where imagination makes miniature toy soldiers into representatives of real human bodies. The crossover between imaginary bodies and real bodies occurs in *Little Wars* when Wells imagines himself as "General H.G.W.," commanding the toy forces as if they were actual soldiers: "His inky fingers become large, manly hands, his drooping scholastic back stiffens, his elbows go out, his etiolated complexion corrugates and darkens, his moustaches increase and grow and spread, and curl up horribly; a large, red scar, a sabre cut, grows lurid over one eye. He expands—all over he expands" (37).

The allure of playing a fantasy war game is here represented as increasing the player's sense of his manliness, expressed in terms of size and rigidity (with obvious phallic connotations). While he was joking about killing the inhabitants of Woking after tea, the fantasy gave him a virtual power over the people around him. Similarly, by imagining himself as "General H.G.W.," he creates an imaginary, battle-scarred, and more masculine version of himself, drawing on stereotyped images of Victorian officers as having curled-up moustaches and ruddy complexions. This is the same allure as playing a board game or a video game, and enacting a more powerful version of the real self in the imaginary space; Wells represents here an early version of the "militarized masculinity" in first-person shooter video games, with the same imaginary investment in a more powerful military alter ego or avatar (Kline, Dyer-Witheford, and de Peteur 247–248). This is also the same imaginative identification with combat by a man portrayed in the *Punch* cartoon reading about and reenacting the charge of the Light Brigade in imaginative identification with the military (see chapter 1).

The imaginary battlefield in *Little Wars* is therefore a space for Wells to perform both his masculinity and his unconscious imperialism. The same factors are at work in the later iterations of fantasy wars, from the board game Risk (1957) through to military strategy video games. The connection between *Little Wars* and later military video games can be traced through Gary Gygax and the creation of Dungeons and Dragons.

Fantasy Wars: Dungeons and Dragons

Gygax was heavily involved in the miniature war gaming communities that were descendants of Wells's *Little Wars*. There were multiple, loosely connected war gaming clubs in the 1960s and 1970s, linked by niche publications such as the gaming company Avalon Hill's *Avalon Hill General* (Peterson 5) and play-by-mail battles, where correspondents would send their game moves to each other and move the units on their maps.[7] Gygax helped bring the disparate groups together by hosting the first Gen Con convention in 1967–1968, which eventually became the largest miniature war gaming convention in the United Sates. He was a fitting successor to Wells because he briefly enlisted in the army, but then became a pacifist later in his life (Peterson 112), showing a conflicted attitude toward real, large-scale violence while embracing fantasy warfare.[8] Gygax also wrote the "Foreword" to an edition of *Little Wars*, in which he proclaimed his indebtedness to Wells. He echoed Wells's defense of imaginary warfare in miniature as an alternative to real wars: "When defending the hobby of playing military miniatures games, I have often quoted or paraphrased Wells' statements—as I do now—regarding the fact that miniature soldiers leave no widows and orphans, and that if more people were busy fighting little wars, they might not be involved in fighting big ones" (Gygax, "Foreword," in Varhola). Gygax acknowledges that reading *Little Wars* was a revelation for him and that it "influenced my development of both the *Chainmail* miniatures rules and the *Dungeons & Dragons* fantasy roleplaying game" (Varhola). He also discusses the effect of World War I on the popularity of *Little Wars,* but does not mention the coincidence of his creation of Dungeons and Dragons with the Vietnam War. Some of the first Gen Con conventions attracted anti-war protestors, which may well have prompted Gygax to cite Wells on "little" wars versus "big" ones because participants in the convention were accused of warmongering (Peterson 111).

In this context, a turn to fantasy gaming in Dungeons and Dragons helped defuse accusations of warmongering because the inclusion of wizards, elves, and dwarfs took gaming into an obviously fantastical setting where the victims were not necessarily human beings. Gygax did not invent the fantasy element out of whole cloth because, as Peterson documents, such elements were already present in miniature gaming with the creation of imaginary countries like Blackmoor (Peterson 66). He and his partner Dave Arneson originally created the country as part of the "Castles and Crusades" group of medieval miniature war gamers, and this landscape was later adapted by Arneson for fantasy characters in *The First Fantasy Campaign* (1977). As *Chainmail* (Gygax and Perren) makes clear, the fantasy element is derived explicitly from J.R.R. Tolkien's *Lord of the Rings* (1954–1955) and includes such characters as Ents and Balrogs. *Chainmail* fused the fantasy characters from Tolkien with the rules developed for miniature medieval war gaming, including turn-taking and tables to resolve combat, and morale checks. Outcomes of battle were all governed by rolls of the dice.

While Gygax and Arneson between them integrated fantasy into miniature war gaming, the central focus of *Chainmail* was still combat. Most of the text was concerned with how much damage can be inflicted by various weapons, such as arrows and swords (11) and gunpowder weapons (13). The rules also included a "burst radius" for damage from explosions, as determined by a die roll, an innovation which Gygax said was derived from Wells's *Little Wars.* This innovation was "an idea that was translated into both the *Chainmail* catapult missile diameters and the areas of effect for Fireballs in D&D" (infinity.net). The "burst radius" calculation is an abstract way of representing the damage to bodies from explosions in real combat, but a die roll dictates how many figures are removed from the battle, because the fire is directed at miniature surrogates for humans that don't have limbs or torsos to register partial damage. This process is apparently bloodless, but it is still an attempt to represent the effects of combat as realistically as possible.

Nathan Shank differentiates play and violence in his discussion of Dungeons and Dragons, citing the presence of rules as key to separating the two terms, and defining the game as "violence as play" (195). Shank's title gestures toward a concept of "productive violence" that oddly never appears in the text of the article, but signals his attempt to recuperate the imaginary violence of Dungeons and Dragons from the actual violence of combat. However, Scarry, in her discussion of the body in pain, sees the crucial factor in war as damage (see Introduction). Scarry's insistence on damage shows how Shank is approaching Dungeons and Dragons as "war without bodies" because his analysis stays at an abstract level in defining "violence." Shank's use of Jacques Derrida (191) also shows that he defines the term as a purely linguistic phenomenon characterized by *différance*. The absence of real bodies purges violence of any material consequences and makes it purely imaginary, but does not address the process of substitution identified by Virilio (see Introduction) by which warfare is derealized.

Wells and Gygax used the distance between fantasy battles and real warfare as a defense of playing war as a game. However, their efforts to make miniature war gaming as realistic as possible make this argument problematic. The problem of realism in miniature figures was evident when the toy manufacturer Britains, Ltd. had to withdraw an exploding trench that it released during World War I. Blowing up British soldiers even in imaginary warfare was a little too close to reality in the context of the real deaths of soldiers in World War I, and after 1918 the company emphasized the production of civilian figures because the market for miniature soldiers was depressed by the aftereffects of the war (Peterson 270).

The relationship between fantasy and real violence is an issue that simmered for the players of miniature war games and Dungeons and Dragons for several years, but it assumed a greater importance when video games became a mass market product involving a much larger consumer base, especially among young men. Whether imaginary violence encouraged players to enact its real-life counterpart was asked much more urgently in this new media environment, but this is too narrow a focus on the issue. The wider issue is whether such games contribute to a wider war culture in which organized violence is made more acceptable. While the critique of imaginary violence is usually focused on first-person shooters, such analysis also applies to the successors of tabletop war gaming in military strategy video games.

Virtual Warriors and Armchair Generals

The immediate successors to *Little Wars* and miniature gaming were turn-based military strategy games that translated many of their conventions into computer code. The creation of games played on computers marked the increasing conflation of war as a game and playing at war as a widespread consumer activity.[9] The first computer-based military strategy games were produced by companies such as TalonSoft Inc. that took the paper and cardboard games of Avalon Hill, in which movement was governed by hexes superimposed on a map, and made them playable on personal computers.[10] TalonSoft was founded in 1995, and in 1998 was acquired by Take-Two Interactive, which helped market the games

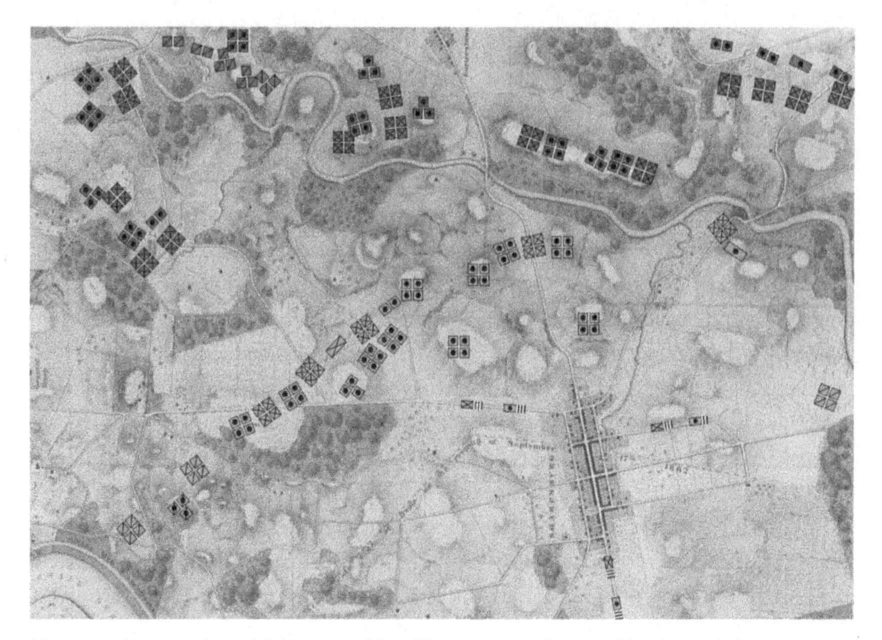

Figure 5. Screen shot of the *General Staff Wargaming System*, "Antietam." Courtesy of D. Ezra Sidran, the *General Staff Wargaming System*.

widely to the rapidly growing personal computer market. The company was dissolved in 2002. Most TalonSoft games re-created historical battles such as Waterloo or Gettysburg and allowed armchair generals to fight the battles over again with the same forces to see if they could achieve a different outcome. In this emphasis on historical re-creation, they followed Wells's interest in older forms of combat; nuclear warfare, for instance, was not an option in TalonSoft games.

This is a somewhat different genealogy from first-person shooter games, which owed more to the video game arcade because they depended on the skill of the player, especially reaction time and dexterity.[11] There has been much research on violence and first-person video games, especially given concern about on-screen violence influencing off-screen behavior.[12] Where first-person shooters simulate interpersonal violence, turn-based military strategy appeal to would-be generals in the style of "General H.G.W." and simulate larger-scale violence between massed units. The violence in these games simulates the role of nation-states in organizing and deploying armies to pursue goals such as territorial expansion. Rather than taking a first-person view of the violence, such games simulate an omniscient, top-down perspective as the player takes control of army units. For example, the *General Staff* game engine gives the player control over tokens representing the different units as if from above in an update of the initial *Kriegsspiel* military simulation (Figure 5).

Military strategy video games, in contrast to the real-time action in first-person shooter games, often allow for contemplation between moves, although speed and reaction times became more important when some games moved into massively multiplayer online formats (MMO) such as *World of Warcraft*.[13] Nonetheless, both types of video games simulate warfare and, while the problem of violence is more abstract in strategy games, they also attempt to re-create combat with some level of similitude. While questions of individual violence are obviously raised more acutely in first-person shooters, real-time strategy games have their own set of issues, because they are heavily implicated in the history of colonialism and replicate Rhodes's dictum that "expansion is everything." Where Wells indirectly reflected the attitudes of the British Empire in *Little Wars*, turn-based strategy games make expansionism and the exploitation of resources the basis of their gameplay.[14]

The *Total War* series of turn-based strategy games released by The Creative Assembly Limited in the U.K. were initially based on historical campaigns but eventually also included more Dungeons and Dragons–like features such as wizards and dwarfs. The first game that the company issued was *Shogun: Total War* in 2000, followed by *Medieval: Total War* (2002), *Rome: Total War* (2004), *Empire: Total War* (2009), *Napoleon: Total War* (2010), *Total War: Attila* (2015), *Total War: Warhammer* (2016), and *Total War Saga: Thrones of Britannia* (2018). The games represent a technological advance over the TalonSoft series thanks to the capability to switch from a strategy level to a battle mode, where the player controls simulations of the bodies of the soldiers in the units, with sound effects to accompany the combat. Titles such as *Thrones of Britannia* show that its main historical base is Great Britain, and the spelling in the game follows British usage. The historical games such as *Rome: Total War* are marketed in terms of spectacle, recalling the allure of descriptions of the charge of the Light Brigade by Russell (see chapter 1), with its emphasis on the aesthetics of battle. The product description for the *Rome: Total War* DVD, for instance, proclaimed that the user would "see exotic ancient cities and colossal armies rendered in incredible detail, as jaw-dropping battles unfold."[15] While this sounds like vicarious tourism, in this case you can see the exotic cities and kill the soldiers defending them.

While the historical setting may vary, the gameplay remains essentially the same. The player must build armies, capture cities and resources, and, above all, expand by conquering other countries. Capturing natural resources such as timber, iron, and gold that are scattered across the playing area gives the player added income and special bonuses for each one, in a video game version of the undeveloped islands in Wells's *Little Wars*. The player must use these resources to upgrade cities and armies while balancing income and expenditure, with heavy penalties for running a deficit. The game rewards both aggression and expansion, based on the premise that in order to thrive, the player's country must be on a permanent war footing to expand its territory and power. In a parallel to the game mechanics of the *Total War* series, Dom Ford has analyzed how *Civilization V* "replicates empire building and the writing of imperial histories and narratives" and reinforces a narrative of domination through technological progress ("eXplore"). As Souvik Mukherjee says, "such a resource-hungry geopolitics . . . creates the binarism of centre and its peripheries" ("Playing Fields" 302), and this binarism replicates the dynamics of both imperialism and Friedrich Ratzel's concept of *Lebensraum*, which was later connected to the expansionist policies of the German National Socialist Party (see Klinke and Bassin). Mukherjee also underscores how playing *Total War: Empire* alters and simplifies the history of nineteenth-century India: "The Creative Assembly version of Indian history, although broadly reflecting the historical geopolitics of the region, is nevertheless happy to simplify the diversity of the region and adapt the map to reflect a comfortably imperialist (and to use Edward Said's concept, "orientalist") set of places, resources and societies" ("Playing Fields" 306). Mukherjee takes this critique further by arguing in "Playing Subaltern" that the games' rules constrain players to follow certain assumptions about their culture, and that players who are marginal to the national identity that the game constructs are unable to protest (511). Mukherjee, playing the game in India, had a different relationship to the power structure implied in the game than I did, playing as a white British male, especially when it came to things like the East India Company and the Mutiny of 1857, when the "enemy" in the game is Indian sepoys. Even calling it a "mutiny" in the title rather than a "rebellion" replicates the imperialist perspective of the British government and the East India Company by labeling it an illegitimate uprising against established imperial authority. As Mukherjee points out, the game's "victory conditions" make conquering Hindustan unavoidable and thus play out the expansion of the British Empire in India; the player must become an imperialist to win. Mukherjee and Emile L. Hammer have applied a postcolonial lens to these games, deconstructing their implicit colonialist biases inherited from the nineteenth century (33). While the replication of imperial attitudes is most blatant in *Empire*, all the Creative Assembly games follow an implicitly imperialist logic that has not changed since Wells's *Little Wars*.

Most studies of video games and violence focus on first-person shooters played by young adults, especially males, and address the effects of the games on aggressive behavior and misogynist attitudes.[16] Peter Mantello, by contrast, addresses the cultural context of the games and the ways in which they align with post-9/11 American foreign policy. Mantello argues that the confluence of military, government, and corporate interests represented in "military shooter" games calls into being modes of understanding and "practices of authorization hardwired into an emerging imperial polity" (486). Like Simons and Lucaites (5), Mantello sees video games as part of a wider culture in which the boundaries between the civilian and military are being blurred (487), as the "MS [Military Shooter] video game injects the logic of production and consumption into a virtual battlefield" (510); this logic also underlies the *Total War* series in its emphasis on seizing resources and expanding the base of production. However, it is not just that the logic of a game mimics the prevailing imperative to accumulate, but that the "video game works in an operational capacity both as ontopolitical primer and interpellative device that calls into being neoliberal subjectivities hardwired into the ethos of a new American militarism" (517). Mantello's analysis suggests that both real-time strategy games and military shooter games interpellate the player to conform to existing regimes of power, and implicitly, and perhaps later explicitly, endorse the same aims and methods employed in the name of "security" and state policy.[17] In other words, the games inculcate governmentality in that the player internalizes state power as part of the natural order and acts accordingly, apparently out of free will, but actually constrained by the "procedural logic" of the computer code.

This is not to say that these games overtly instantiate such subjectivity, but rather that the game mechanics themselves create a certain structure and logic. Mantello quotes Ian Bogost on the rhetoric of video games, saying that they "argue with processes rather than content" (489), so that the subject position of the gamer stays the same while the historical context of *Total War* campaigns changes. While I have been writing about video games in general and *Total War* in particular in a typically objective and scholarly way, I must now confess that I play these games and was playing *Total War: Warcraft* while I was writing this chapter. In conclusion, I will therefore address directly and personally one of the aspects left out of the analysis so far, and that is the pleasure derived from playing simulations of war in military strategy games.

The Pleasures of Conquest

I'm an enthusiastic war game player, but also consider myself a pacifist, which is obviously contradictory. I could claim, like Wells and Gygax, that such play is purely imaginary and better than actual war, but such a defense is untenable in the context of the militarization of American society and the confluence of violence and entertainment in video games. As David Leonard says, "the blur between real and the fantastically imagined, given the hyper-presence of war on television and within video games, constructs a war without bloodshed, carnage, or destruction" ("Unsettling"), or in other words a simulation of war without damage to bodies across different media. C. Richard King and David J. Leonard make a similar point in describing "the absence of blood" in video games, which they link to the way in which, thanks to the sanitized coverage on news networks like CNN, "the absolute horrors of war are lost on many people" (101). These analyses underscore that in playing these games I am, like Wells in *Little Wars*, reproducing the wider political context, and in its procedural rhetoric normalizing wide-scale violence.

The games that I play are apparently "war without bodies," but they turn "flesh and blood adversaries into screen-mediated targets of opportunity" (Mantello 487), with the result that playing at war reinforces the hegemony of the military-industrial entertainment complex, even if the "enemy" is nonhuman creatures such as those in *Total War: Warhammer*. The game mechanics remain a powerful indoctrinator that must be addressed to combat what Leonard calls the "marked failure to recognize video games as sophisticated vehicles inhabiting and disseminating ideologies of hegemony" ("Unsettling"). The imaginary spaces of military strategy games "affirm the political and cultural status quo from which they originate: They reproduce, charge, and disseminate interpretations, ideologies, and worldviews in contemporary society" (Riegler 60), especially the underlying assumption that violence is inherent in human nature; in this they replicate the ideology of Victorian military masculinity (see chapter 1) while updating it for the asymmetrical U.S. "war on terror."

While I am critical of Shank's premise of "productive violence," his emphasis on rules and structure captures some of the appeal of the games when he writes that "the rules of a game and the rules of a juridical system both create space and elasticity in the system which allow for a freedom of movement within boundaries" (185); to play Dungeons and Dragons or a video game is to enter a zone with rules that demarcate the range of choices. The threats within the game follow a clear set of procedures, unlike in the non-game environment, and the responses are equally clear and limited. The game space is a comforting environment because it is rule-based and provides solace for the precarity of actual existence in the twenty-first-century United States. However, the "freedom" that Shanks cites is, of course, illusory because the boundaries are determined by the game mechanics, which carry their own ideological message. The term hegemony applies here because to play the game is to voluntarily surrender freedom of movement for a sense of security purchased at the cost of consent to a militarized approach to solving problems. Michel Foucault's use of "power" is germane here, in that power does not simply say "no" but also encourages a subject position that aligns with the prevailing model of obedient citizenship.[18] When I play these games, I assume exactly the kind of neoliberal subjectivity that Mantello has critiqued.

The pleasure in assuming this subject position derives from conquest; this is the experience of "ludic war" that Matthew Thomas Payne analyzes in the context of first-person shooters, but also applies to the strategy games that I play (11). When I play these games, even if I am a race of lizard "men" (women are in the minority in the ranks of most armies in the game), these are "my" troops and they are defeating an enemy and increasing "my" territory. Once a region is occupied, the map is "redrawn and carries your nation's colour" (Mukherjee, "Playing Fields" 301) in a replication of imperial expansion, and there is an undeniable pleasure in looking at a map that reflects the growing extent of one's own territory. On the wall of my office I have a map showing the expansion of the British Empire across the centuries, and a timeline of the expansion of my lizard man empire is a virtual analogue to that historical document. This was particularly problematic when I, as a British male, was re-creating the British Empire in *Empire: Total War*, including the conquest of India. Unlike Mukherjee, I could identify with the invading forces while being uncomfortably aware that I was also re-creating a history of oppression and exploitation, even if only virtually.

The problematic issues with the *Total War* series are not solved entirely by moving into the fantasy world of *Warhammer*. Apart from the game mechanics and thus the procedural rhetoric being the same as previous versions, the first "race" that I played was Lothern, a nation in the game's parlance of "high elves." There are some women in the Lothern campaign, but the elves are exclusively white and their enemies are "dark" elves. While "light" and "dark" are primarily meant as references to good and evil, the vocabulary can easily shade into racial stereotypes based on skin color. The description of high elves on the Creative Assembly *Total War: Warhammer* site sounds very much like Wells's fantasy topography, because the "island kingdom" of Ulthuan is engaged in warfare against "lesser races" (Creative Assembly), and their resources are there for the taking. The term "lesser races" recalls the Victorian attitude toward the subjects of the empire, summed up in Rudyard Kipling's "The White Man's Burden," which advised the U.S. to take over the role from the British Empire of governing "Your new-caught, sullen peoples,/Half devil and half child" (*Stories and Poems* 359). This parallel with British imperialism is reinforced by one continent being a rough analogue of Africa, which was, of course, dubbed the "Dark Continent" in the Victorian era. The map of *Warhammer*, like that of Africa in the imperialist imaginary, needs to be explored, subjugated, and incorporated into an expanding Lothern empire. The other continent in the game is a rough analogue to South America and is populated by Lizardmen, who seem heavily influenced by Aztec architecture. Overall, the "races" in the video game "function as hegemonic fantasy by filtering the racial imagery," thus making "whiteness" a default setting (Higgin 3) and excluding identities that reflect the diversity of the gaming population.[19] Not only do the scenarios in the game replicate imperialist attitudes, but the video game industry itself is "enabled by a historical and material global network dependent on the imperialist capitalist system across the world" (Hammar and Woodcock 55), so that the exploitative economics of the industry is reflected in the structure of the game.

War games are a subset of competitive play, and the overall pleasure of playing such games lies in winning. At the end of a campaign the software congratulates you on vanquishing your enemies, as well as celebrating your "achievements" in battle as the game progresses. Achieving goals such as the number of enemies killed gives some satisfaction, but the pleasure I derive also aligns me with a culture of military-style games that support the idea of violence as a means to an end, and the idea that a war has clear "winners" and "losers" in a simple binary. Philip Beidler, who has spent a career meditating on the aftereffects of war, points out that the violence has not "ended" for millions of Vietnamese who still feel the effects of herbicides sprayed on the countryside (161), and Rob Nixon in *Slow Violence* (2011) documents the many ways in which the environment in Iraq has been poisoned. While the games that I play are "wars without bodies," they are still imaginary violence and join war movies and other cultural products that present such conflicts as free of material consequences and ending in unambiguous victory (in marked contrast to the Crimean and Iraq wars). To counter the absence of bodies in games like *America's Army*, Joseph DeLappe would type in the names of soldiers killed in the Iraq War as he played because "this game exists as a metaphor, not wanting us to see the carnage, the coffins coming home. It's been sanitized for us" (Craig). Just as with news coverage of the Iraq War in the United States, images in video games are sanitized by excluding the true carnage.

Video games and war became more intimately intertwined in the first Gulf War in 1991–1992 and the subsequent war against Iraq in 2003. The conflict was extensively documented through video footage, and included images of buildings being blown up by drone strikes. Viewing virtual soldiers on the screen as an experience of war without actual bodies in the early twenty-first century became a literal "war without bodies" as media coverage excised images of the dead. Instead, war was represented as psychological trauma, and the emphasis was placed on therapeutic interventions with the establishment of the diagnosis of post-traumatic stress syndrome (PTSD).

CHAPTER 4

Trauma and the Soldier's Body

During the Iraq War American media agreed not to show images of dead American troops or report Iraqi civilian casualties, thus making the conflict what John Taylor in *Body Horror* (1998) has termed "a war apparently without bodies" (157).[1] On the other hand, American media coverage frequently represented returning American soldiers as psychologically traumatized, and reported that they had suffered severe mental injury from which they had to be cured.[2] The diagnosis of post-traumatic stress disorder (PTSD) in particular was applied to returning soldiers and defined the aftermath of combat primarily as a wound to the mind rather than a physical injury.[3] Where Scarry in *The Body in Pain* (1985) defined war as damaging primarily to the body (63–67), American media discourse about war has increasingly represented it as having lingering psychological repercussions that in the context of PTSD are exhibited as an inability to forget the past. The emphasis on war as psychological trauma in the diagnosis of PTSD was paralleled by the repression of bodily pain in media coverage; the effects of war were defined as a mental disorder that needed to be cured rather than physical death and destruction. PTSD was initially a diagnosis of the psychological symptoms of veterans of the Vietnam War, but was applied to soldiers returning from Iraq as part of the representation of "war without bodies." PTSD has subsequently become a pervasive term applied to a variety of traumatic experiences beyond war.

I use the term "trauma" in this chapter to denote the direct experience of violence that causes so much distress for a subject that he or she repeatedly returns to the memory of the event (although not necessarily involuntarily). Trauma is different from "shock" in that it infers emotional repercussions long after the event rather than a sudden, discrete jolt. I do not question that people experience trauma, but rather challenge how it has become a catchall term that obscures wider power relations in its focus on individual recuperation. My definition differs radically from such broad uses of "trauma" as that of E. Ann Kaplan in *Trauma Culture* (2005), who uses the rubric "mediatized trauma" to include those who witness events through the media or hear of events from other people. According to Kaplan, viewers are "traumatized" (2) even though they only see representations or hear secondhand accounts of events. She asserts that extending the definition to include experiencing terror "should not make the term meaningless," (1), but the imperative "should not" portrays an underlying anxiety that this expanded usage is indeed veering into incoherence. Applying this definition to war would imply that someone could be traumatized by watching footage of combat as much as someone who experienced the violence directly, which indicates how diffuse the term has become.

Autobiographical texts written by those who served in Iraq challenge the diagnosis of the trauma of war as a mental disorder and reinscribe the body in the war experience. Memory itself is pathologized in the diagnosis of PTSD, which turns soldiers' reactions to the violence of war into damage to the mind rather than an emotional and spiritual response to extreme violence. Cathy Caruth characterizes trauma as a "crying wound" and toggles between damage to the mind and the body in her use of the term (*Unclaimed* 8). Treatment of PTSD also treats memory as a "wound of the mind," and focuses on mental damage, implicitly excluding the soldiers' body. The treatment of PTSD stresses the inability to forget events that intrude on everyday life. Memoirs by soldiers deliberately relive trauma as a way of coming to terms with the experience rather than seeking to forget what befell them.[4] First-person texts about war voluntarily return to this wound and deliberately give it voice; they mobilize memory to combat trauma rather than trying to forget the experience. Such texts also represent the bodily experience of soldiers that was expunged in the sanitized coverage of war in the American media.

Caruth, taking her cue from Freud and psychoanalysis, persistently characterizes trauma as "inextricably tied up with belatedness and incomprehensibility" (*Unclaimed* 92), and a great deal of her analysis is forensic, aiming to diagnose the repressed or hidden origins of traumatic memories. The texts I discuss in this chapter are different because the writers have no trouble remembering the events they describe, in keeping with Raymond McNally's findings in *Remembering Trauma* (2003), where he concludes that there is no evidence of narrative fragmentation resulting from trauma (135), and that difficulty remembering events by subjects in studies of trauma "concerned everyday forgetfulness, not amnesia from the trauma itself" (190); indeed, those who attempted to repress their memories were unsuccessful. Far from having difficulty accessing their memories, veteran authors from the Iraq War recount the traumatic events in their autobiographies with no problems of amnesia or incomprehensibility. Building on McNally, Joshua Pederson has argued that "critics should turn their focus from gaps in the text to the text itself" and turn away from a model of "traumatic amnesia" (338). Pederson also notes that narrating trauma has restorative value (338–339).[5]

Authors of autobiographical texts from the Iraq War seek to convey the experience of being both the bearer and the victim of violence in a combat zone. Their dual focus contradicts the assumption in the diagnosis of PTSD that sufferers of trauma are positioned solely as victims; the subject position of a soldier is much more complex than this simple model would suggest. As Caruth herself has remarked, PTSD represents a challenge to the diagnosis of trauma itself:

> On the one hand, this classification and its attendant official acknowledgment of a pathology has provided a category of diagnosis so powerful that it has seemed to engulf everything around it. . . . On the other hand, this powerful new tool has provided anything but a solid explanation of disease: indeed, the impact of trauma as a concept and a category, if it has helped diagnosis, has done so only at the cost of a fundamental disruption of our received modes of understanding and of cure, and a challenge to our very conception of what constitutes a pathology. (*Trauma* 3)

Caruth is correct that PTSD presents a challenge to the categorization of trauma as a "pathology," and this challenge is echoed by Iraq War veterans; however, she errs in naming PTSD in terms of "disease." This definition of trauma as a pathology in PTSD needs to be revised as it applies to soldiers, because it medicalizes what may well be a humane reaction to violence. Texts by those who have experienced combat in Iraq do not represent trauma as pathology, but rather as witnessing or as a recording of acts of violence to which they have been subject and incidents in which they have been complicit.[6] They collectively represent an attempt to make others understand the contradictory relationship of the soldier to violence and utilize memory as testimony to the suffering caused by war for them and others. These writers are what Michael Rothberg has termed "implicated subjects," because they are "neither simply the perpetrator nor the victim though potentially either or both at other moments" (xv), so that the binary of perpetrator/victim in PTSD is contradicted by their autobiographical narratives. These writers are both victim and perpetrator, and they experience trauma not only as the subject but also as the source of violence as "implicated subjects."

Caruth, however, also points to another problem with trauma itself as a category, namely the way in which the scope of the diagnosis has broadened (*Trauma* 4), or what Luckhurst describes as the "transition from medical discourse to a cultural condition" as the word has spread throughout popular culture as a way of understanding the effects of violence (*Trauma* 62). PTSD is now a diagnosis that can be applied to a wide range of traumatizing events, including major stress, sexual assault, natural disasters, and warfare. The catalogue of causes of trauma shows the extent to which it has become a term that erases the context of the event in favor of a homogenizing label. Like the term "war" itself, trauma has become associated with a wide range of disparate causes, diluting its original focus, which was on the mental turmoil of veterans of the American war in Vietnam.

Didier Fassin and Richard Rechtman in *The Empire of Trauma* (2009) have argued that "trauma has become a general way of expressing the suffering of contemporary society, whether the events it derives from are individual (rape, torture, illness) or collective (genocide, war, disaster)" (19–20). The category of trauma has become overgeneralized as a label that is affixed to an array of incompatible events that efface important differences between them. The term "trauma" in PTSD encompasses so many different scenarios of violence that it occludes important distinctions in scale and the relationship between perpetrator and victim, so that genocide is now placed in the same category as assault, thus erasing the distinction between individual versus collective violence. Such a wide range of scenarios for trauma implies a loss of scale in determining its effects. Similarly, McNally has challenged the finding "that as many as 89.6% of American adults were trauma survivors eligible to be assessed for PTSD" as an index of how much the category has been expanded and made overly broad ("Expanding Empire" 46). Clearly, there is a problem in terminology with a diagnosis that labels the majority of Americans as traumatized, and this widening scope of "trauma" as a category needs to be questioned. The authors I examine here show that, *contra* the diagnosis of PTSD as involuntary memory, they react to the causal traumatic events by voluntarily returning to them to make sense of the past through autobiographical narratives.

The Soldier's Gendered Body

In addition to the trauma of combat, female soldiers in Iraq faced the possibility of sexual assault by other members of the military. For them a "war without bodies" was impossible because they are persistently defined in terms of their gender and sexuality, and subject to the danger of sexual assault. The Department of Defense releases an annual report from its Sexual Assault and Response Office (SAPRO) to document the number of reported incidents; in 2016 it released its eleventh report on the number of cases and steps taken to address the problem.[7] As the SAPRO report made clear, both men and women suffer from sexual assault, although men are more reluctant to report such incidents because they will be viewed as "less than a man" (Garago). Women in the military are at a higher risk of sexual assault than men (Lamothe) and consequently have higher rates of PTSD, according to Lutwak (359). For female soldiers, their situation is complex because they find themselves both empowered by being in the military and disempowered by sexual assault by male soldiers.[8]

While they are in a uniform that overtly makes women equal with men by supposedly erasing individual differences, sexual assault places women as potentially subject to violence and reasserts wider gender dynamics. This conflict between the authority given by the position of "soldier" and the effect of being victimized in sexual assault is most powerfully and graphically illustrated by Kayla Williams's *Love My Rifle More Than You* (2005); in this autobiographical text she portrays an "implicated subject," occupying subject positions as empowered thanks to her status as a soldier with authority over civilians, and as herself the object of violence when she is sexually assaulted by a member of her unit. As an implicated subject in Iraq, she was both the possible source of violence as a soldier with a gun and a victim of violence when a man tried to coerce her into sex. When Williams was threatened by a fellow soldier named Rivers, her contradictory subject position as a woman and soldier was brought into conflict. Finding himself alone with her one night, Rivers tried to force Williams to masturbate him by grabbing her arm and pulling her hand toward his crotch, keeping a grip on her even as she tried to resist. She dissuaded him from continuing his assault by reminding him of his girlfriend and telling him she was not interested, but he continued to try to pull her toward him (207). Although Williams was able to defuse the situation, the threat of sexual violence subverted her identity as a soldier.

The incident immediately evoked conflicted feelings for Williams, who throughout her career in the army wanted to prove that she was a soldier on equal footing with the men in her unit and did not want to call out for help: "The *shame* of being in a position where you might have to do that. Yell for help. Like some damn damsel in distress. Knowing that you would have to explain what just happened here" (208; italics in original). In common with many survivors of sexual assault, Williams felt both shame and anger at her assailant, and had trouble dealing with these contradictory emotions. She found the feeling of helplessness at that moment unnerving because it undermined her image of herself as a powerful and autonomous female soldier. She also rejected the label of "some damn damsel in distress" as disempowering, although in this moment Rivers asserted his power over her by refusing to let her go. She rejected the stereotype of the helpless female, but her identity as a soldier was threatened by this encounter because she was being defined as a sexual target and in terms of her body. Rivers was not alone in undermining her by focusing on her body in this way; other men in her unit referred to her as "boobs" (167), objectifying her in terms of her physical difference from their masculinized identity of "soldier."

Williams's experience shows that the uniform does not eradicate the body in the military, underlining that treatment based on physical differences persists despite attempts to integrate women into combat units.[9] The increased presence of women in combat units that were heretofore solely male has added the complications of heterosexual sexuality in a unit. Soldiers are not yet robots (see

Conclusion), and they may well find themselves sexually attracted to each other, but the gender dynamic in Williams's case put her in a vulnerable position and potentially subject to force. Reporting the assault would have threatened her relationship with the other men in her unit; she writes that "I have to assume that if it comes right down to it, the guys would all back him. As somebody in their team, in their unit, in their MOS.[10] One of the boys" (208). Williams was obviously not "one of the boys," so Rivers would be believed over her if she were to lodge a complaint about the incident. Williams was confronted through this incident with the masculine bias in military culture and the way in which she could be ostracized based on her gender.

Williams was also aware that, while the military ostensibly had a mechanism for reporting sexual assault, making use of the process would provoke animosity from fellow soldiers:

> As much as the Army would like to tell us that it's not true, girls who file EO (equal opportunity) complaints are treated badly. . . . Needless to say guys do not like girls who file EO complaints. They will talk shit about them. They will not want to be around them more than is absolutely necessary. . . . Even *girls* don't like girls who file EO complaints—they don't want to rock the boat. Girls don't want to be perceived as filing a frivolous complaint. There's still the assumption that girls lie about harassment to get what they want—to advance their careers or punish somebody they dislike. (209; italics in original)

While there was gender-based solidarity among the men, this did not extend to women, who viewed any complaints filed by another woman "as rocking the boat" and causing unnecessary trouble in a culture that valued order and obedience. She and others like her kept quiet about sexual assault thanks to the toxic mixture of shame and the opprobrium that accompanies women making such accusations. Williams herself uses condescending gendered language in writing about this incident; she refers to "girls" who file complaints rather than women, which implicitly infantilizes them. In the subtitle of her book, she identifies herself as "young and female," whereas male soldiers writing about their experiences in Iraq don't feel the need to specify their age or their gender in their titles, indicating once again that the term "soldier" is marked as male.[11]

The incident sent Williams into a deep depression as she tried to come to terms with the conflicting emotions caused by the assault. Her reaction to such harassment is typical and parallels the case study in Alicia Rossiter and Sharlene Smith's diagnosis of PTSD in female veterans (364–366).[12] In order to conform, Williams had to adapt to misogynistic attitudes herself, joining the other men in the unit, for instance, in telling rape jokes (168–169). While Williams would like to see Rivers's assault as an anomaly, the conversations that she reports elsewhere in her text suggest very thinly veiled violence against women implicit in military culture:

"What gets you hot?" It's Matt to me.
"Small dicks. Unmanly men." I look around. "You guys."
"Screw you, hatchet wound."
This talk gives me a nasty shiver. (167)

Williams does not make the connection in her book that jokes about rape and assault made it difficult for her to report sexual assaults by normalizing a culture of sexual violence against women. Williams wanted to be "one of the boys" by making jokes about male genitalia just as they do, and to file a sexual assault charge would be to place herself outside the fraternity and foreground her female identity. Although she would like to, Williams could not avoid being defined in terms of her body's difference from the men around her, as the epithet "hatchet wound" attests. This repellent epithet both defined her in terms of her genitalia and made her the object of assault because it foregrounded both her body and her potential vulnerability to violence. It is little wonder, therefore, that this conversation gave her a "nasty shiver," as it positioned her as the subject of sexual violence and presaged her later attempted assault by Rivers. After the incident Rivers circulated a false story that Williams herself had initiated the encounter and, as she feared, she found herself excluded from the fraternity of her unit in terms of her sexuality and body, "And now the guys I considered my friends were treating me like a *girl*. I was tits, a piece of ass, a bitch or a slut or whatever, but never really a *person*. *Bros before hos*" (214; italics in original). Williams shows how "personhood" resides in the masculine body, and that to be defined as a "girl" (once again infantilizing her) placed her outside the male brotherhood. After the incident she is defined in terms of her body and its difference from the phallocentric discourse of the male soldiers, which excluded her from an organization that expressed relationships on the basis of the male anatomy. As other autobiographical texts from the Iraq War show, male soldiers' language is phallocentric, so that women in the military have to adopt masculine linguistic norms and deny their own bodies to be accepted, as Williams did in telling sexist jokes. The male physique is the implicit norm in "soldier," and women are defined by their difference, making their participation in war one defined by their bodies.

Jason Christopher Hartley's *Just Another Soldier* (2006) gives many examples of how this phallocentric discourse operates, making his and other male soldiers' bodies central to his text. One of Hartley's favorite expressions in his memoir is "brokedick" (15), and early in the text he relates an incident in which one soldier put his penis instead of a wrench in another soldier's hand. Hartley says of this incident, "the jury is still out on who is the more gay, the guy who touched the dick or the guy who let a guy touch his dick" (25). While this incident raises questions about homoeroticism in the military, it also shows how the language and actions in his unit were based on male anatomy, especially their genitals. Hartley also makes the centrality of the male body clear in his argument that women should not be allowed in the military because of menstruation, pregnancy, and breasts, all of which for him mark them as different and inferior in his masculine terms (95).

There was therefore no subject position for Williams as a woman simply to be a "soldier," and she had to continually struggle to find a space within military masculine discourse because of the "social roles imposed on women" through sexuality (Irigaray 186–187).[13] This is one aspect of the "invisible war" that female soldiers face in addition to combat. As Williams says: "A woman soldier has to toughen herself up. Not just for the enemy, for battle, or for death. I mean toughen herself up to spend months awash in a sea of nervy, hyped-up guys who, when they're not thinking about getting killed, are thinking about getting laid. Their eyes on you all the time, your breasts, your ass" (13). Williams portrays in her text a highly sexualized environment where, when fear of death does not dominate their thoughts, soldiers are thinking about sex, and any woman in the vicinity becomes the focus of the male gaze.[14] There were long periods of boredom for the troops, which would mean there was very little to distract them from sexual reveries. Women faced tension not only from being in a combat zone but also in having to fight what Williams terms "a separate bloodless war within the larger one" (22).

Williams is a self-aware and astute writer who in her memoir also addresses the ways in which attention from men can be seductive and affirm the power of her sexuality, "Still, it's more complicated than that. . . . Their eyes, their hunger: yes, they're shaming—but it also makes you special. I don't like to say it—it cuts inside—but the attention, the admiration, the need; they make you powerful. If you're a woman in the Army, it doesn't matter so much about your looks. What counts is that you are female" (14). Williams resisted being defined in terms of her body, but she also recounts some of the power that comes from being "Queen for a Year" (18) in the army because simply being female guarantees receiving attention that affirms your attractiveness to men. Of course, this means that Williams is being defined again in terms of her body and sexuality, and she registers her unease at the situation with a metaphor of damage to her body, which is "cut inside" at this admission (14). Williams also admits that because of this attention the members of her platoon "become *your* guys" and jealousy sets in if another woman comes into the tent because "these guys, they're your husband, they're your father, your brother, your lover—your life" (14). These possessive feelings made Rivers's attempted sexual assault doubly traumatic because it broke the idea that the men in her platoon were "her" guys. Where male writers express bonds in the military in terms of brotherhood, Williams adds husband/father/lover in a catalogue of male roles that are obviously not all compatible with sexuality, underlining the conflicted position of a woman entering such a patriarchal institution, or what Williams calls "the most authoritarian institution imaginable" (14) and "a massive frat party" (21). The assault by

Rivers disrupts these bonds by excluding her from the "frat party" of the U.S. military and breaking the bonds with the men in her unit.

Williams's text shows why it is impossible to categorize female soldiers using the current definition of PTSD because of their complex and contradictory relationship to trauma; she is an "implicated subject" because, as a member of the military, her rifle is a symbol of power, but at the same time she is the potential subject of sexual assault and thus disempowered. Williams's book title quotes a line from a Marine Corps song about loving a rifle, and the cover shows her holding a weapon to which she professes attachment, saying that "I do love my M-4" (15). The rifle represents her power and access to the identity of soldier, but also underscores the phallic identification of men with their weapons. Drew Cameron's "Living without Nikki" (2008) is not about a woman, but rather his M16A2 rifle that he named because the weapon was for him and other soldiers "our companion, an extension of our bodies" (80). His weapon was therefore for him a symbolic female presence: "Sleeping with the strap wrapped around my arm I would nestle her against my side for extra comfort some nights. . . . She became part of me, I was part of her. One of the first things I did when I returned was give her up. They took Nikki away and locked her up" (80). The repeated use of "her" and "she" and the female name make the rifle into a surrogate girlfriend.

If a rifle is an extension of your body, no longer having it in your life is both a separation and an amputation. Williams did not profess this level of attachment to her rifle in her text or give it a name, because she did not have the same access to such phallocentric discourse. While she may have loved her weapon, she did not describe it as an extension of her body. However, she did find herself in positions of authority thanks to her weapon, such as when she had to deal with a difficult detainee, who she cursed at. As she says: "I don't like to admit it, but I enjoyed having power over this guy. . . . I was uncomfortable with the feelings of pleasure at his discomfort, but I still had them. It did occur to me that I was seeing a part of myself I would never have seen otherwise" (205). Williams sees this incident as a contrast to the feeling of powerlessness that all soldiers felt in Iraq, and that she felt when threatened by Rivers, but this was a situation in which she had power over a man, whereas in her relationship with the men in her unit she had to negotiate a quite different gender dynamic (206).

Williams was acutely aware of the dangers of having absolute power over another human being, and she experienced this firsthand when she was asked to help with an interrogation. She was a witness to incidents that made her uncomfortable, such as mocking and humiliating the prisoner in sexual terms (247), and acts of violence that contravened the Geneva Conventions (246). Williams in her text links these events to those at Abu Ghraib prison in 2003, where photographs of American soldiers humiliating and torturing Iraqi prisoners caused a scandal in the United States (250–251). The images of Iraqis being humiliated and tortured put civilian bodies squarely into the frame of American media coverage and temporarily disrupted the easy flow of images of a "war without bodies" (Eisenman 7–8).[15]

Williams toggles in her book immediately from a situation of potential perpetrator of violence as a guard to victim in the assault by Rivers; this juxtaposition shows how a simple definition of her as a victim is impossible because she occupies two contradictory subject positions simultaneously. While Williams says "I don't forget. I can't forget any of it" (17), this is not because she is suffering from PTSD but because she deliberately revisits these episodes and writes about them to tease out the complex psychological issues.[16] Her thoughtful exploration of the gender dynamics that set female soldiers apart in their experience of violence makes her autobiography one of the most compelling texts to emerge from the Iraq occupation. Her memoir shows how the identity of "soldier" was predicated on the male body and is disrupted by the presence of women's bodies; her text shows a "war with bodies" in Iraq. Her ambition to face unflinchingly the contradictions in her identity as a female soldier echoes texts by writers who participate in the "Warrior Writers" initiative who want to, in Brian Turner's words, "open the hurt locker" (15) and voluntarily reexperience trauma.[17]

PTSD and Moral Injury

While Williams tackled the gender binaries created for the female soldier, texts from the "Warrior Writers" project address the soldier/enemy distinction that underlay the attitudes toward the people they were fighting in Iraq. In order to kill another human being without remorse, you have to dehumanize them, often through derogatory terms, and to make sure "that we didn't see our enemies as people" (Williams 200). In Drew Cameron's "You Are Not My Enemy," he tries to counter the dehumanization of Iraqis as "hajis" and the objects of violence:

> You are not my enemy
> My brother my sister,
> But I have done something wrong
> And perhaps I am now yours. (3)

Like Williams, Cameron makes an effort to understand the Iraqi perspective, and he rejects the label "enemy" that is applied to them wholesale; he complicates his relationship to them as a soldier by replacing enmity with familial relationships that imply that he should be protecting them. Cameron humanizes the people with whom he came into contact in Iraq. Similarly, Jack Lewis's "Road Work" describes his vehicle killing a civilian and then how he found a link with the victim's father through their shared grief as parents, since he himself had lost a child (123–126). In both cases the authors' works describe their attempts to establish human links with the Iraqi civilians with whom they came into contact. This kind of narrative questions the simplistic branding of an "enemy" in Iraq, where the boundaries between innocent civilians and hostile forces was not clear.

Cameron clearly feels a great deal of guilt over his actions, from having "done something wrong," and the poem is motivated by guilt and a need to atone for past actions as a soldier. "Guilt" is not a word that appears in the DSM-5 criteria for PTSD, but it is an emotion that is prevalent in the writing found throughout autobiographical texts from the Iraq War. Caruth addresses guilt only once in *Unclaimed Experience* in relation to Freud's views on Judaism and Christianity (18). Guilt, however, is central to the memories of many veterans. Many writers feel culpable for actions ranging from the treatment of Iraqi civilians to murder, and they write about their actions as a form of atonement once back in civilian life. Expressions of guilt are found throughout the writings of Iraq veterans, both in the "Warrior Writers" program and in the text and video found in *Operation Homecoming* (Carroll). By expressing guilt, these writers make Iraqi deaths grievable and put noncombatants back into the frame. The effect of guilt in these writings is a textual equivalent to that of the visual images that I discuss in the next chapter, which reinsert the civilian body into the frame of narratives of war.

The personal essays in *Operation Homecoming*, edited by Andrew Carroll, are a collection of nuanced and subtle explorations of the experience of war coming out of the Iraq conflict. It was supported by the National Endowment for the Arts and released as both as a book and a documentary commissioned by the cable channel Home Box Office (HBO).[18] The contradictory emotions of a soldier facing combat are captured in Denis Prior's "Distant Thunder," found in both the *Operation Homecoming* text and the documentary. His essay opens in Kuwait before the opening offensive of the Gulf conflict, and he writes that "I hate the idea of war and I can't wait for it to begin," conveying conflicting emotions about the prospect of war (25). He experiences a similar divided consciousness when witnessing the bombardment during the "shock and awe" phase of the offensive:

> The night the sky lights up with artillery. We are shelling Hillah, we are shelling Karbala, we are shelling Baghdad. Since we paused the artillery has caught up with us, so after dark we watch the streaks of fire thrusting up into the sky, and listen for the cool free fall back down to earth, and then see the flash, then the boom, as it pummels cities, like lightning, then the thunder of a rainstorm. The shelling cleaves me in two, one side shaken, knowing each flash and boom means more innocent Iraqis dying in their homes, the other side stilled, knowing it also means less of the enemy likely to shoot at me. I sit and watch, picturing the Fedayeen and Republican Guard getting annihilated. Die, motherfuckers, die, I say to them all. You. Not me. (36)

Prior describes being "cleaved in two" by his warring feelings about the artillery bombardment of Iraqi cities, which he knows may be killing Iraqi civilians.

Like Cameron, he expresses a complicated relationship to violence that acknowledges both the enemy and the death of civilians. Where the death of civilians was repressed in American media coverage of the war, Prior acknowledges that the violence unleashed in war is killing noncombatants as well as opposing troops and makes their loss grievable. He gives a striking portrait of his divided consciousness as a soldier and his position as an "implicated subject." While witnessing the shelling of Baghdad, American cable channels were showing video of the "shock and awe bombing of the city." In such live video there were explosions but no casualties visible, making it an apparent "war without bodies." Prior's narrative undermines the position of spectator by acknowledging the deaths that were hidden from view.

Prior's reaction here is best described through the concept of "moral injury," which refers to the "direct participation in acts of combat, such as killing or harming others, or indirect acts, such as witnessing death or dying, failing to prevent immoral acts of others, or giving or receiving orders that are perceived as gross moral violations" (Maguen and Litz). As Rita Brock and Gabriela Lettini explain, "moral injury is not PTSD" (xii), and their approach emphasizes constructing a coherent narrative that is "the result of reflection of memories of war" (xiv). This is precisely the method followed by veterans such as Price, who manage to come to terms with their actions during the Iraq War; "moral injury" thus provides a framework with which to approach memories of war and guilt at actions taken without suggesting that such reactions are the result of a "disorder," as implied by PTSD. Matt Howard addresses the same issue in his rejection of PTSD as a "disorder" when he asks "what part of being emotionally and spiritually affected by gross violence is disorder?" (163). The approach of "moral injury" also places emphasis on the importance of autobiographical narratives in coming to terms with the past, underlining the therapeutic aspect of writing about feelings of guilt at past actions in the Iraq war.

While these writers express guilt at being the bearer of violence, they were at the same time also potentially the victims of violence at the hands of Iraqi fighters opposing the occupation of Iraq by foreign troops.[19] The violence in this case was not sexual assault, as in the case of Williams, but the daily possibility of being shot at or blown up, meaning that soldiers had to live daily with the possibility that they might be killed. This fear placed them in a contradictory subject position that underlies many of the autobiographical poems from the Iraq conflict. Brian Turner's poem "Here, Bullet" expresses this dual subject position best; the poem is part of the collection *Here, Bullet* (2007), which is the most complex and nuanced book of poetry to emerge from the Iraq conflict. The poem is addressed to a bullet that is personified and seems to be invited to strike the body of the soldier. On first reading, the poem seems to describe the impact of a bullet on his body, but as the poem develops, his body itself also becomes a rifle:

Here is where I moan
The barrel's cold esophagus, triggering
My tongue's explosives for the rifling I have inside
Of me. (13)

Where Williams represented the subject position of a woman through her body, Turner embodies the ever-present possibility of physical damage and death in his poem. Turner's poem represents war through the body. The perspective shifts in his poem from inviting the bullet to strike his body, to the speaker's body itself having a "rifled" throat; his esophagus has spiral grooves that make the bullet rotate, thus giving it greater accuracy over long distances. The shot could still come from the unnamed enemy firing the gun, although human agency has been erased by personifying the bullet itself in a consistent displacement of human desire onto the weapon. The emphasis on moaning and the line "inside of me," however, makes it more likely that the narrator is firing here, and that the poem itself mimics both firing a gun and the action of a bullet striking Turner's body. By giving the bullet agency as a projectile launched from the narrator's throat, the poem uses Turner's own violence to counter the threat of the incoming bullet. Even though the poem is couched as an invitation to the bullet to strike him, it also preempts and tries to negate presaged injury by becoming a source of violence itself. In different terms than in Williams, Turner's poem registers the position of a soldier as both perpetrator and victim of violence. Turner, like Cameron with his gun "Nikki," as a man can more easily identify his body with a weapon than Williams, but he registers the same sense of precarity as Williams because of the threat of violence.

In an interview in the *Operation Homecoming* documentary, after reading "Here, Bullet" aloud, Turner says that he carried this poem around with him while he was in Iraq (Richard Robbins 1:00–1:05). The poem in one sense was a talisman that, by imagining the worst, seeks to ward off death; it also, however, pre-imagines trauma. The poem foregrounds bodily trauma by describing "the clavicle-snapped wish,/the aorta's opened valve" (13) as the bullet hits its target. The poem imagines both killing and being killed simultaneously and expresses the soldier's omnipresent fear of death:

Each twist of the round
Spun deeper, because here, bullet,
Here is where the world ends, every time. (13)

The final "here, bullet" shifts from imagining the body as the permanent site of trauma to death as the "world ends." The "every time" in the last line invokes trauma as repetition and shows how, even if a soldier's body is not damaged, the psychological effects remain in the omnipresent threat of death. Living with the omnipresent fear of attack makes dying almost a relief, because imagining being killed every day, and carrying the poem around with you, imagines an end to the uncertainty. "Here, Bullet" is an imagined suicide, a welcoming of the destruction of the body as a relief from the omnipresent sense of an impending death. However, writing these fears down in a poem is also a way of working through the pre-imagined trauma and, like the other texts examined here, it is a voluntary exploration of the damage caused by violence in war. Turner strikingly expresses the trauma of war through his body in "Here, Bullet," reversing the trend of making war a primarily psychological wound. "Here, Bullet" makes the soldier's vulnerable body visible and represents the precarity that images of heroism often repress and that the American military would like to banish through automation (see Conclusion).

Tim O'Brien in the *Operation Homecoming* video says about memories of war that "I think there's a false notion that we all ought to recover from everything. . . . I believe the opposite. I believe that there are some things you should not heal from, they are unhealable and even if they are healable you oughtn't to do it anyway; there's something to be said for remembering and not healing" (Richard Robbins 2:00–2:25). Using the vocabulary of physical healing, O'Brien counters the prevailing model of PTSD by insisting that remembering trauma is not a disease but a humane response to war. He asserts that the emphasis should not be on healing the "crying wound" but on remembering and coming to terms with the events that caused trauma. The same sentiment is expressed in Jon Turner's letter reprinted in *Warrior Writers* expressing regret over his actions as a soldier in Iraq. He suggests that he needs to preserve these memories because he does not consider them to be the result of a disorder: "due to what I have experienced and witnessed, I now have post-traumatic stress disorder. I'm sorry but I do not consider this a disorder. Due to my actions that I willingly did, it is my fault that I have flashbacks. . . . I did not sign up to see innocent people get hurt and die" (Jon Turner 119). His text directly criticizes the labeling of memory as a "disorder" in PTSD, arguing that war is a trauma that should not be forgotten. Where the clinical approach to PTSD emphasizes healing, these narratives carry out a process of witnessing and remembering. The conditions of war are such that feelings of guilt and responsibility may be a human and humane response to unbearable experiences in combat situations. The concept of "moral injury," while continuing the language of mental wounds, suggests that PTSD is a human response to the violence of war and not a disorder.

Clearly, if a veteran is suffering debilitating effects from trauma, then therapy is the appropriate response, but these texts suggest that for many writers deliberately remembering trauma can itself be therapeutic. Clinical studies have shown that autobiographical "scripts" provoke the most intense emotional reactions in veterans (Pitman et al., quoted in McNally, *Remembering Trauma* 119), and writing autobiographical texts for these writers is the most appropriate response to trauma. As Brian Turner says of his poetry, "I may not be a very good soldier, but I may be a very good witness," so that his poems are a record of his time as a soldier in a combat zone (Richard Robbins 1:10–1:13). His texts and those by other veterans bear witness to the trauma of war and insist that the lessons that the writers have learned not be forgotten. If trauma is a "speaking wound," as Caruth says, then these autobiographical texts assert that the wound should be allowed to continue to speak as testimony to the effects of violence on both the instigator and the victim.

The Politics of PTSD

By focusing on PTSD in soldiers returning from the conflict in Iraq, the media framed the conflict as a question of the individual psychological reactions of soldiers to extreme situations where their lives were threatened, but without reference to the policies that placed them in danger in the first place. Wider issues of power and the political causes of warfare are excluded from this frame. This is not to say that concern over the psychological impact of war on soldiers is misplaced, but rather that the focus on psychological trauma restricts the conversation about the costs of war; as Rothberg has argued, attention needs to be paid to the "systems of violence" that surround a traumatic event (xiv), and this is especially urgent in the case of war. The focus on the psychological impact of war on soldiers made the loss of civilian life "ungrievable" (see Introduction).

The focus on soldiers as psychological victims of war fused "traumatic social memory" and "individual testimony to horror" (Fassin and Rechtman 93), so that the healing of soldiers came to represent also a recovery from the collective memory of war for American society, just as the body of the physically wounded soldier served a similar function in the aftermath of the Crimean War (see chapter 2). The discourse of PTSD surreptitiously emptied the trauma of war of "all reference to the collective history" (Fassin and Rechtman 282) and absolved the wider society of any responsibility for the promulgation of war. Allan Young has argued that PTSD is embedded in discourse "by the practices, technologies and narratives with which it is diagnosed, studied, treated, and represented and by the various interests, institutions, and moral arguments that mobilized these efforts and resources" (5); it is a social construct, and its use often effaces the political context in which it is deployed. This political effect during the Iraq War obscured the wider power dynamics that placed soldiers in such a contradictory position in Iraq in the first place and effaced the toll of war on both soldiers and civilians. PTSD as a discourse reinforced war culture in the way it was deployed and made death in the Iraq War ungrievable by excluding bodies from the frame.

Autobiographical narratives by veterans of the Iraq War frequently address this political context, especially those written for the "Warrior Writers" project by authors such as Fernando and Maria Braga in "Food, Water & Revolution" (122). Such texts describe the impossible situation in which they were placed when, as soldiers, they were supposed to be "liberating" Iraq and protecting civilians, but often inadvertently killed innocent people. They name the political forces that deployed them and lament their role in inflicting violence on others. By speaking as "implicated subjects" who are both the instruments of violence and wielded by wider political forces beyond their control, they represent the emotional complexity of war, a complexity that is lost under the current rubric of PTSD.

Just as autobiographical narratives by soldiers about the Iraq War reinsert the body into the frame of war, visual artists and authors of fictional texts also counter the loss of bodies and of political content in the media coverage of conflicts. Most notable are the photographs of Sophie Ristelhueber and Gohar Dashti, who each in different ways attempt to document the cost of war through imagery. Where Ristelhueber focuses on the post-conflict signs of damage to the landscape and alludes indirectly to the bodies that were subjected to violence, Dashti deliberately places the civilian body in battle scenes in her photographs. In a reaction against media coverage that frames a "war without bodies," they attempt to disrupt conventional narratives and imagery that make warfare acceptable.

CHAPTER 5

Sophie Ristelhueber

LANDSCAPE AS BODY

French photographer Sophie Ristelhueber takes photographs of landscapes damaged by violence. These landscapes record the aftereffects of conflict without showing either the perpetrators or victims, alluding to deaths that must be imagined but are not represented. One image from *Eleven Blowups* (Figure 6) in particular seems to recapitulate Fenton's *Valley of the Shadow of Death* photograph (Figure 1); it shows a nondescript valley with a crater in the road. The only sign of the preceding violence in the empty landscape is the damage to the road. To create these images, Ristelhueber looked at "video rushes from Iraq, taken by local Reuters correspondents" and her own photographs to document the ongoing cycle of war and destruction in the Middle East (*Operations* 382). She used scenes from the videos and combined them with her own photographs to create the images, so that they are "real" but assembled from different times and places. In *Eleven Blowups,* she subverts the documentary power of photojournalism by creating images in which "everything is true and false at the same time" (*Operations* 382) to evoke endemic violence from specific incidents that are part of a longer history of such acts, or what Brian Turner in *Here, Bullet* called "an echo of history, recited again" (1). Both Ristelhueber and Turner in different ways capture the recurrence of war, not to suggest its inevitability but to convey the sense of narratives repeated in different epochs and countries that echo across history.

While Ristelhueber's photograph of a crater in a valley would seem to echo Fenton's cannonballs in the Crimean War, the media environment for her image was radically different from that of Fenton's nineteenth-century photograph. Fenton's photographs reinforced the image of heroic British masculinity, whereas Ristelhueber's photographs circulate in a media environment saturated with sanitized accounts of warfare that eschew depicting dead bodies. Ristelhueber's images of a "war without bodies" are designed to unsettle the conventional media

Figure 6. Sophie Ristelhueber, *Eleven Blowups #1*, 2006. Digital print, 110 × 133 cm.

reporting of conflicts by using the landscape as a surrogate body that registers deaths that we cannot see but must infer from the traces left behind.

Ristelhueber also took a series of photographs of the Kuwaiti desert after the first Gulf War and published a selection of them in *Fait: Koweit 1991* (1992). Ristelhueber's photographs depict the effects of war on the landscape of Kuwait, but, like Fenton, Ristelhueber decided not to photograph dead bodies. They evoke what Jean Baudrillard termed a "blank war, or a war even more inhuman because it is without human losses" (308) in their images of a desert landscape with no human figures. No combatants are visible in any of her photographs, only the aftermath of war such as spent munitions and craters. Ristelhueber said of her images, "the absence of man in my images reinforced his presence," making an argument for the power of an image of war that excludes bodies (Stauble). For Ristelhueber, representing war without bodies was the most effective way of dramatizing the damage caused by human violence. Retracing the flight of Iraqi soldiers northward on the ground, like Fenton retracing the charge of the Light Brigade, she came across "a collection of shaving brushes, razors, and little mirrors that must have formed part of the soldiers' kits" (*Operations* 280), and she recorded these abandoned objects as relics of a human presence in the desert. She was also drawn to the marks left by bomb impacts that drew on her previous work on "scarred territories" (Hindry 75). The craters and trenches left after the war evoke for Ristelhueber a wounded body, and the scars left on the earth are like those on human skin (Mellor 222).

Whereas documenting war through photography was a novelty in the Crimean campaign, Ristelhueber published her photographs when the norms of photojournalism against which she is reacting were well established. The dominant aesthetic for Ristelhueber is not the Victorian ideal of "beauty," as it was for Fenton in such images as *The Valley of the Ribble and Pendle Hill* (Baldwin plate 78), but rather a visual record of damage to the landscape from war or other acts of violence as testimony. When Fenton photographed a landscape, he chose scenes that were informed by an idyllic image of rural England (see chapter 1), and after the Crimean War he became known as "the greatest master of landscape photography in Britain" thanks to his panoramic views of the countryside (Gernsheim and Gernsheim 29). The cannonballs that litter the valley have not damaged the scenery, and if they were not present his photograph would just be of a road in a valley. Ristelhueber, however, shows quite dramatic damage to the landscape, and the point of her photograph is to document the destructive aftermath of war through wounds in the earth.

Where Fenton photographed cannonballs resting on the surface of the earth, Ristelhueber is drawn to scarring in which the earth is incised and left with visible scarring. Ristelhueber used *Operations* (2009) as the title of a book that brought together landscape images from *Fait* with photographs of postsurgical scars from *Every One* (1994). Ristelheuber herself has said that "for me, the bodies and the territories are the same thing" (quoted in Gustafsson 76), and she records the effects on both with surgical detachment. *Operations*, the title of one of her published works, means both military maneuvers and surgical incisions. One image from *Every One* in particular, that of a woman's back with a scar running the length of her spine (*Every One* #14), recalls images from *Fait* of traces of trenches and roads on a desert landscape (figure 7). The same interest in trauma recorded as incision and scars informed her photographs of postsurgical human bodies. As Hindry says, the desert is "like a living skin furrowed by wounds that are visually experienced in pain" (76). The landscape is therefore a surrogate human body, and in the case of *Fait* represents not just any bodies but those of absent soldiers. The term "operations" hovers ambiguously between the surgical and the military, just as the landscape in *Fait* shows damage that suggests the absent soldier's body mutilated by war.

Ristelhueber summons the absent human body with images of trauma in the landscape rather than as the "wound of the mind" of PTSD (see chapter 4). Ristelhueber's post-battle images are a visual corollary to PTSD, showing how the effects of violence linger long after combat has ceased. Rather than focusing

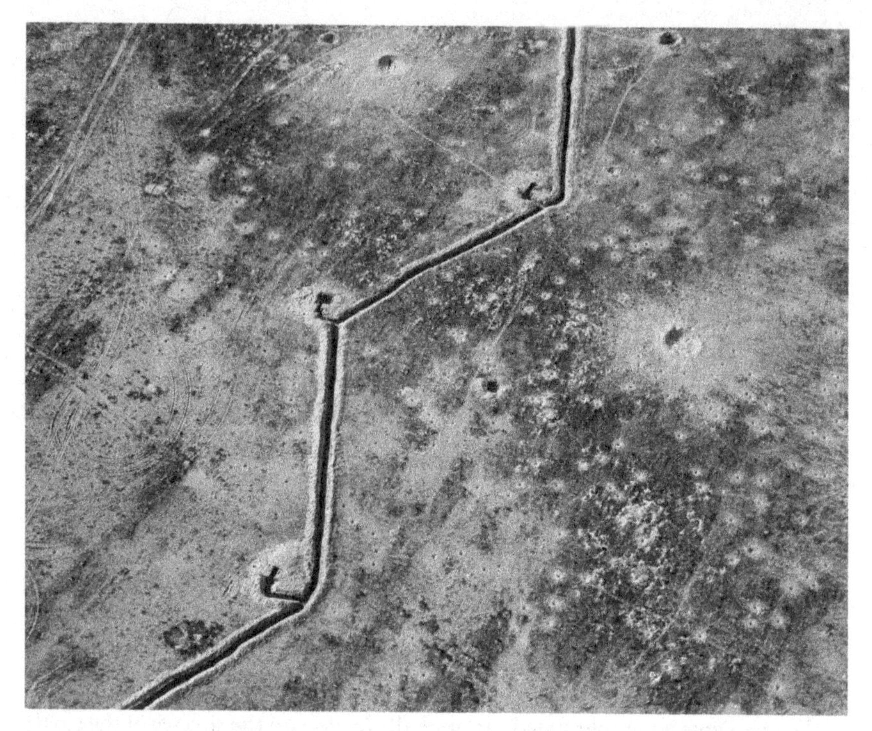

Figure 7. Sophie Ristelhueber, *Fait #20*, 1992. Chromogenic print, 100 × 127 cm.

directly on the moments in which damage is inflicted on a body or a landscape, she captures the aftermath and the scarring that preserves an afterimage of violence or violation of the earth.

Ristelhueber's images are poised ambiguously between photojournalism and art because they combine "two apparently incompatible modes of looking at a site" (Vandermeulen and Veys 28), forensic and aesthetic; she documents damage to the landscape in studiously distanced terms, but her photographs are also informed by an artistic sensibility. David Campbell has analyzed how war photography has been framed since Robert Capa's images of the Spanish Civil War, arguing that "the conventions of war photojournalism have been frequently aligned with the state" (29).[1] Ristelhueber's images are not aligned with the state, although members of the U.S. military helped fly her over Kuwait to take her photographs (*Operations* 281); she sees herself as more of a photographer using images for artistic purposes than advocating a position vis-*à*-vis state-sponsored violence (Hindry 53). Her photographs of the Kuwait desert are witnesses to the destructive power of war rather than photojournalism, but they are nonetheless framed by the conventions of the genre in the way that Campbell suggests, in that they are "based on the premise that power is held accountable by making visible a different picture of war than that offered by the state" (27). While Ristelhueber undoubtedly counters conventional images of war, her Kuwait photographs do not diverge from contemporary representations of war because of the wider context in which they are produced. Assumptions about warfare encoded in other media influence how her images are perceived, especially in the focus on material damage rather than on bodies.

Ristelhueber's images are embedded in a media environment in which soldiers' bodies and those of civilians are rendered invisible for very different reasons. Her photographs participate in a process that Virilio traces to World War I and the use of aerial photography, and the beginning of "a growing derealization of military engagement" and weaponizing of images of war (*War* 1) (see Introduction). They were also published in an environment where such photographs "contribute to creating a culture in which war is paradoxically in/visible" (Simons and Lucaites 5); that is, images of war are both ubiquitous and unremarkable and pass unnoticed. Butler, in *Frames of War*, has characterized this as the way in which images "organize visual experience" so as to determine "what can be seen and what can be heard" (3). Ristelhueber excludes the original scene of violence so that her images will have greater emotional resonance, but the media context for her images is one that inundates viewers with images of war, but also evacuates them of evidence of bodily damage. Such images, especially in video footage from the front lines, played an important role in the first Gulf War of 1991–1992 and the subsequent reinvasion of Iraq in 2003.

Fait and Drone Vision

Ristelhueber's aerial images of landscape in *Fait* overlapped with a view of the landscape shot from a drone; she took photographs from a plane flying over the Kuwait desert, and she conveys a distanced perspective on the effects of war. She said in an interview that "for me it was as if I was fifty centimeters from my subject, like an operating theatre. I was so sure about what I wanted to do" (quoted in Hindry 76). She uses a "surgical" vocabulary based on a hospital operating room, and in doing so echoes the vocabulary of "surgical" strikes used to describe "smart" weapons like drones (Vågnes 8). In her photographs the aftereffects of war are seen "from above," not at the ground-level perspective of those engaged in or suffering from the effects of combat, employing the aesthetic of "aftermath instead of immediacy; the trace instead of the event; distance instead of proximity; withdrawal instead of immersion" that Bruno Vandermeulen and Danny Veys see as the sign of the effect of digital media on photography (28). The absence of bodies further replicates what David Hastings Dunn has termed "disembodied aerial warfare" (1237). Dunn is referring to the lack of human operatives in drones, but the phrase also captures the absence of human bodies in the coverage of both the Gulf and Iraq wars. One image in particular resembles a small town surrounded by bomb craters (*Fait* plate 43). This is actually a military installation, but the afterimage of human activity contrasts eerily with the absence of bodies, either living or dead. Ristelhueber is right that the inferred violence visible in the damaged landscape is powerful, but like Fenton's image of cannonballs in a nondescript valley, the message is equivocal; is the damage to the landscape or the atrocity of human on human violence being memorialized in the photograph? Clearly for Ristelhueber it is both, but other aerial images of such damage promote the idea that "surgical strikes" by drones avoid killing noncombatants.[2]

As Butler points out, "the aerial view" is often preferred to graphic photos of dead soldiers or maimed children because it is "established and maintained by state power" (*Precarious Life* 149) and reinforces existing power structures. Ristelhueber uses aerial images for a purpose diametrically opposed to the representation of state power, but her choice of aerial views also echoes the way that the U.S. military, in alliance with major news corporations, dictated the norms under which images were consumed and interpreted by American viewers. Aesthetic images by artists countering the normalization of war are swamped by the proliferation of implicitly pro-war propaganda that normalizes unending conflict and makes state-sponsored violence unremarkable. Aerial views are just one visual trope in this process, especially images shot at a distance from drone strikes.

The photograph of the crater in figure 6 is itself a product of video imagery. As she said, she looked at video rushes from correspondents in Iraq and selected images to combine with her previous photographs. There has been much debate over whether Fenton composed the scene in the *Valley of the Shadow of Death*, and the photograph of the crater by Ristelhueber is equally the product of arrangement by the photographer. However, any photograph is framed by the camera or by later editing, and these two images are only more extreme examples of a selection process. While her image is based on video, it is recast as a photograph and invites pause and reflection on the violence that preceded the cannonballs in the valley or the crater in the road. Her photographs are intended as what Henrik Gustafsson terms a "cut" or intervention "into the circulation of images of conflict perpetuated by the military and the media alike" (70), and are meant to disrupt the easy consumption of war imagery by raising an ethical imperative (71).

Where Fenton exhibited his photographs as part of a new technology for representing war, Ristelhueber's images coincide with mechanical views of war from cameras in drones. There has been much debate about whether drones represent a new technology of warfare (see Dunn; Greene; Holmqvist), but the new military hardware is radical visually in inaugurating what Stahl terms "a new way of seeing" ("What the Drone Saw" 661). The process initially described by Virilio in *War and Cinema* has evolved from "video-missiles" (1) into self-propelled drones equipped with telescopic lenses and missiles. In his insightful article, Stahl defines drones as a medium for visual images rather than just a weapons system, and argues that they constitute a particular way of both representing warfare and of framing the responses to the deaths of those viewed through the drone's lens, or what he terms "a mosaic of visual discourse—what I call, for the purposes of this article, 'drone vision'" (659). Daniel Greene characterizes this as "seeing like a drone," which is symptomatic of "a new, imperial visual culture of war" that he argues is distinct from previous disciplinary regimes (234). So ubiquitous have drones and drone imagery become that they have expanded from an "until now predominantly military—scopic regime into everyday life" (Weber) and are thus unremarkable. Stahl and Greene have both analyzed the alliance between visual targeting and the military in the same way that Virilio outlined the close development of photography and then film as an integral part of warfare. The term "drone vision" is a powerful way to articulate that drone footage itself has become a way of understanding war, a way that makes warfare not a question of human bodies but of material "targets" of "surgical" missile strikes. The frame of the video screen in drone vision occludes damage to human bodies.

"Drone vision" is found in media reports, YouTube videos, and video games that convey a vicarious experience of warfare via video. In some ways this is nothing new; the paterfamilias discussed in Chapter 1 who was reading about the charge of the Light Brigade in the newspaper was using print media as a virtual experience. The quantity and quality of images, however, has increased since the Victorian era, and the first Gulf War was a "video war" in many senses. Not only was it the first war with direct video feeds from the front lines; it also took place in the context of many other representations of the conflict, as did Fenton's photographs of the Crimean War. As with Fenton's photographs, this framing helped shape the civilian response to the conflict and set a pattern for the ensuing invasion of Iraq. Commenting on coverage of the Gulf wars, Cynthia King and Paul Lester write that "the 'frame' of media gathering and packaging works in conjunction with a reader's 'frame' of how she has learned about the world through previous media reports" (626), showing how this synergy between media and consumer works. What cannot fit within this "frame," including civilian casualties, goes unreported.

What seemed novel about the first Gulf War and its successors was the sense of "virtual immediacy," thanks to technological advances such as "drone vision" and embedded reporting, "A 'real-time' war conducted and viewed at high speed, a 'smart' war projected through myriad networks onto linked screens in the bomber's cockpit, on TV, and the internet, giving the potent sensation of instant access, and total, continuous immersion" (Nixon 202). Nixon's formulation underscores how the combination of military hardware in "smart" weapons (possessing software that supposedly enables them to make intelligent decisions), media coverage on television, the diffusion of images and videos through the internet, and the gamification of warfare (see chapter 3) all reinforced a visual representation of war. This combination gave a sense of immediacy to the events, but this is true of any new technology. The telegraph, for instance, was hailed when it was invented as a medium that would shrink the globe and make communication immediate.[3] Whatever the technology, however, the experience of noncombatants not in the war zone will be vicarious, because it is always mediated by print or visual media. Stephen Prince calls modern consumers of such images "voyeurs of warfare," as they "expect warfare to be great drama" (235–256), but this judgment is a little harsh because war was viewed as a spectacle during the Crimean War as well, so the expectation that reports of war could be entertaining is not new (see chapter 1). The desire for drama cannot itself be labeled voyeurism, and the blame for problems with the representation of war lies with producers of media rather than consumers who are conditioned to view war as entertainment. Contemporary audiences are socialized to expect certain narrative structures, and these structures frame war as a drama in ways that encode the expected response. The problem lies with the frame that implicitly privileges the perspective of the military and makes the death of both soldiers and civilians ungrievable.

"Drone vision" focuses the viewer's attention on the power of military technology rather than the human toll of warfare and makes it a "'spectator sport,' in which nothing very meaningful, or very 'real,' seems to be at stake" because of the absence of death (Carruthers 241). The media's self-censorship in agreeing not to show the dead bodies of American soldiers was only one aspect of this lacuna. Phillip Knightley argues that the successful news management by the government resulted in the view of a "war almost without death" (5). Michelle Kendrick has also noted the way in which the emphasis on drones and other technologies mechanized war coverage and erased the soldier's body by replacing it with a "collectivized robot" that "disembodies the killed body twice over" (292). The focus on "smart" weapons and their precision masks both the death of the civilians who are often the unacknowledged victims and the recognition that war still involves soldiers whose bodies are damaged. However, the current trajectory suggests that the military is moving toward a total "war without bodies" as it mechanizes all aspects of warfare, as I argue in the Conclusion.

Michael Ignatieff has drawn the line connecting media coverage and a "war without bodies": "A war apparently without bodies is an imaginative and bureaucratic feat achieved by direct omission (as in censorship), by metonymic transfer onto objects such as machines, and by the media's adherence to polite discourse when reporting state killing on an unknown scale" (157). This description recalls Ristelhueber's use of the landscape and the detritus of combat to indirectly evoke violence, but where she wishes to use the scarring left by war to represent past violence, Ignatieff analyzes an attempt to deny death by excising bodies from the frame. While their aims are completely different, both procedures rely on displacement, on the one hand to foreground antiseptic images of mechanized warfare that apparently involves no human bodies, and on the other the use of images of the effect of war on the landscape to document the damage. Antonio Monegal has expressed this as "picturing absence" in Ristelhueber's images, because she is "resorting to an apparently dehumanized abstraction" and "a form of indirection, the vestiges of violence evoke what is no longer there" (261). Monegal is correct in his analysis, but his use of "dehumanized" draws attention to some of the strategic problems of such imagery in an environment where media coverage generally eschews showing human bodies. This is not to argue, however, that direct representation of damaged bodies alone would have a different effect; Ernst Friedrich's *Krieg dem Kriege!* (*War against War!* 1924) documented in horrific detail the mutilation of soldiers' bodies in the hope that such images would preclude any further wars. The failure of this visual text to change attitudes to warfare underscores the power of large-scale propaganda to mobilize populations in favor of war despite the experience of previous generations.

Perhaps the most insidious frame for drone vision is gamification, which is part of a longer historical process of making war into play (see chapter 3). While gamification is seen as a benign trend in commerce and education, in the context of war it contributes to the derealization of warfare. The first Gulf War in 1991 was dubbed the "video game war" (Stahl, "Digital War" 145). Ignatieff adds that "the bombing of Baghdad was the first war as light show and the aerial bombardment of Iraqi forces was the first battle turned into a video-arcade game" (168). Prior's ambivalent feelings at the bombing of Baghdad, which I discussed in the previous chapter, showed an alternative, humane view of the video. Ignatieff's reference to a "video-arcade game" highlights the trajectory of warfare from film to video arcades to home entertainment in video games. James Der Derian has dubbed video games a component of the military-industrial-media-entertainment network (xi), which blurs the line between the home and the battlefield. Der Derian cites the adoption of *Doom* as a component of training by the Marine Corps (89), and Ahmad has noted that "the military has essayed to recruit gamers as potential drone operators" (26), so that playing at war at home has become a free training ground for potential soldiers, as well as a pastime for noncombatants sitting in front of their screens.

Video games do not necessarily promote war culture. Harun Farocki created a simulation of a battle in the Iraq War as a "serious game" that was demonstrated as a potential treatment for PTSD; as he says, "never has war been so transparent, so tangible, so efficient or so virtual" (125). His re-creation of war was meant as therapy rather than reinforcing war culture. However, while his software was designed as therapeutic, his description of war as "virtual" underscores how ubiquitous war imagery has become thanks to digital technology. As

I argued in the Introduction, technology is not monovocal, and battle simulations can be used for therapeutic goals such as the treatment of PTSD, but the majority of military strategy games and first-person shooters reinforce a militarist and expansionist mindset (see chapter 3).

Drone vision itself has become part of video game culture, Stahl argues, because "the drone has gained an increasing presence in video games themselves, including the two most successful franchises to date: *Battlefield* and *Call of Duty* (665). The "War Is Fun" photo essay by Nina Berman demonstrates how the ubiquity of drone vision contributes to "in/visible war" through integration into daily civilian life (Simons and Lucaites 111–124). It is the "fun" aspect to which Berman gestures that is perhaps the most insidious aspect of how drone vision is framed. Caroline Holmqvist sees the way in which video games draw the viewer into an imaginary environment as part of their appeal: "it is the immersive quality of video games, their power to draw players into their virtual worlds, that make them potent—this is precisely why they are used in pre-deployment training" (26). Video games and movies are both immersive and connected to "fun" (much like the pleasure in strategy games that I discussed in chapter 3), so that when they are linked to drone vision the imagery is evacuated of death and damage to bodies and made into entertainment.

Ristelhueber's images are designed to jolt the viewer out of "drone vision" by forcing an imaginative reconstruction of the violence that preceded them. They are meant as an intervention into the steady diet of derealized violence that most people consume. They gain power not only from their indirect representation of war, but also from their chronicling of another aspect of the first Gulf War, namely the environmental damage to a fragile desert ecosystem. Where Ristelhueber takes an objective, distanced view of the landscape and combat, first-person narratives by American veterans of the Iraq conflict register the effects on a visceral level.

Landscape and the Soldier's Body

When Fenton photographed landscapes in the Victorian era, he usually chose bucolic English scenes (Baldwin 44) and complained about the lack of aesthetic appeal in the flat Crimean landscape (Keller 129) as he adapted the new medium to documenting the Crimean War (see chapter 1). Ristelhueber's images from 1991 register another shift in photographing war from Fenton's time period, and that is an ecological awareness of the aftermath of war.[4] This was particularly acute after the invasion of Kuwait, when the retreating Iraqi army torched oil wells that spewed toxic contaminants across the desert. Industrialized warfare destroys both the built and the natural environment, but the first Gulf War took place in a time of heightened awareness of ecological degradation, thanks to groups like Greenpeace. Ristelhueber's images reflect a growing awareness of and concern for the effect of humans on the environment in what has been termed the Anthropocene.

Nixon has documented the long-term ecological damage from the first Gulf War, which is another aspect excluded by the framing of war coverage: "The Gulf War offers a dramatic instance of how challenging it can be to narrate the ecology of the aftermath because America's corporate media represented the war as a spectacular achievement of speed and untainted victory—a strategically, technologically and ethically decisive war, the nation's anti-Vietnam" (200).

The American military used depleted uranium as a weapon in the Gulf and Iraq wars, and it will remain in the landscape for an unimaginable amount of time because "depleted uranium has a half-life of 4.51 billion years" (Nixon 201). While the war may be "over," the damage to the environment lingers well after hostilities have ceased. This is now recognized with the aftereffects of Agent Orange in Vietnam and the lingering damage from land mines from that and other conflicts, but depleted uranium represents an exponential increase in environmental damage.

Ristelhueber's focus on the fissures in the landscape of Kuwait dramatizes a kind of damage that is hard to imagine in other terms when documenting what has been called "the most toxic war in Western military history" (quoted in Nixon 204). The problem facing any artist or journalist is how to represent a time scale of damage lasting millions of years, represented by such weapons as depleted uranium, which so far exceeds human experience. Ristelhueber's invocation of the human body as an analogue to the landscape is indirect and allusive; first-person narratives of war by veterans, in contrast, reinsert the soldier's body into the landscape and dramatize the experience of war as a visceral experience. As with Kayla Williams in the previous chapter, embodiment is central in memoirs by soldiers.

Anthony Swofford's *Jarhead* (2003) traces the route of an American soldier traveling at ground level over the same terrain that Ristelhueber photographed for *Fait*. In contrast to Ristelhueber's detached and aesthetic appreciation of the landscape as a body, Swofford experiences oppression and suffering in the Kuwait desert: "This is the pain of the landscape, worse than the heat, worse than the flies—there is no getting out of the land. No stopping. After only six weeks of deployment the desert is in us, one particle at a time. . . . Sand has invaded my body: ears and eyes and nose and mouth, ass crack and piss hole. The desert is everywhere" (15). Swofford itemizes in graphic detail the effect of the landscape on the soldier's body and the experience of the desert as an alien invader. Sand finds its way into places where it should not be, making the desert inescapable on a granular level. The landscape also cannot be escaped because it is so monotonous in its "dead repetition" (135). Whereas Ristelhueber's images turn things like trenches into aesthetic objects, for Swofford and his fellow soldiers, digging them is an exasperating experience; they must contend with what he calls "this *most unstable material or medium that will make futile all effort or endeavor*" (177; italics in original) because it thwarts their efforts to create stable walls. Swofford experiences the landscape as frustration and discomfort, and he rejects any attempt to see a redeeming aspect in the trenches because "no one will confuse the outline of our defensive position with the ancient tradition of scarring the earth for the benefit of the gods" (178). Where the "scarring" of the landscape by trenches for Ristelhueber is a sign of past violence, for Swofford trenches represent a mundane and onerous task undertaken for protection from artillery. The damage he and his fellow soldiers inflict on the landscape has no symbolic significance or redeeming qualities for them.

Swofford, however, sees beauty in artillery as it explodes in the sand: "The rounds explode beautifully, and the desert opens like a flower, a flower of sand. As the rounds impact they make a sound of exhalation, as though air is being forced out of the earth" (189). This is the "before" for Ristelhueber's "after" images of craters in the desert landscape. Where Ristelhueber documents the patterns left by such shells, Swofford uses organic imagery to convey witnessing the explosion ("like a flower") and, in an echo of Ristelhueber, sees the desert as a body ("a sound of exhalation"). Swofford also sees an aesthetic appeal in the craters left by artillery in a textual equivalent to Ristelhueber's photographs: "the sand, where the rockets have impacted—smoke-shrapnel burnt into the desert—looks like an abstract charcoal portrait, broad strokes beginning and ending nowhere" (197). For Ristelhueber a crater appears "as if the earth was sucked out from its center" (*Operations* 380), emphasizing absence, whereas Swofford uses the analogy of painting. Swofford, by invoking art, looks at war through an aesthetic lens and recalls Fenton's painterly approach to the Crimean landscape.

Ristelhueber's image of the crater was itself inspired by "a magazine picture taken from a Jaguar aircraft which showed the black impact of bombs which had exploded on the ground" (Mellor 226). This image made her "obsessed by the notion of a desert that had ceased to be a desert" and so she started a project that she saw as similar to her earlier work on "cicatrized territories" (226). The reference to a "desert that had ceased to be a desert" suggests a combined interest in the scars of violence as well as the ecological effect on the landscape. Swofford's text similarly records the widespread ecological damage in his descriptions of the "petrol rain" that fell on the desert after the Kuwaiti oil fields were set alight (214). He and his fellow soldiers entered "a burning, fiery oil hell" as they journeyed through the smoke and fallout from the oil well fires (200). Strangely, after enduring this oily rain for several days, Swofford opened his mouth and tried to taste it, saying that "the crude tastes like the earth, like foul dirt, the dense core of something I'll never understand" (214). Others in his company warned him of the dangers of ingesting pollution, but after being coated in the substance for days, he seemed to want to understand and come to terms with oil pollution by incorporating it into his body. His ingesting the oil from the fires created a physical bond with the desert landscape that was being polluted just like his body. This was a visceral, bodily experience of the ecological damage of war.

Whereas Ristelhueber's images eschew the dead bodies of Iraqi soldiers, Swofford came across many in Kuwait, and he describes them in gruesome detail, especially in one horrific tableau:

> On the other side of the rise, bodies and vehicles are everywhere. . . . I am looking at an exhibit in a war museum. Two large bomb depressions on either side of the circle of vehicles look like the marks a fist would make in a block of clay. A few men are dead in the cabs of the trucks, and the hatch of one troop carrier is open, bodies on bodies inside it. The men around the fire are bent forward at the waist, sitting dead on large aluminum boxes. The corpses are badly burned and decaying, and when the wind shifts up the rise, I smell and taste their death, like a moist rotten sponge shoved into my mouth. (224)

Swofford sounds like Fenton when he retraced the route of the Light Brigade in its aftermath and saw skeletons (see chapter 1); just as Fenton's diary recorded the bodies in the landscape that were outside the fame of his photographs, so Swofford's text gives an account of the immediate aftermath of violence that was excluded from media coverage. His reaction to dead bodies, however, was detached, because in his memoir he compares them to an "exhibit in a war museum" (225). This implicitly made him a spectator, and therefore not directly responsible for the dead bodies that he is viewing. The "presence of so much death," however, made him realize that "I may never again be so alive" (225). Swofford's differing reactions are typical of soldiers' divided feelings about death, as shown in chapter 4; while they feel guilty at witnessing the killing of other people, there is also relief because this means that they have a better chance of survival because of the death of enemy soldiers. The description of the dead bodies also widens the frame of coverage by reporting such a gruesome scene, describing some of the horror of war that was noticeably absent from media coverage that represented a "war without bodies." Swofford's description captures what John Taylor, in the title of his book, evocatively calls "body horror."

The overall effect of memoirs by Swofford, Prior, and others is also to widen the frame to include both the soldier's body and death into representations of war. The dead bodies are usually those of Iraqis, and the memoirs are by American soldiers who survived, but apart from texts like Lewis's "Road Work," discussed in chapter 4, they do not often register the death of civilians. There are, however, visual artists who incorporate the civilian body into images of combat in order to highlight the "collateral" damage of warfare. Their images counter the representation of "war without bodies" by including noncombatants threatened by military violence.

Reinserting the Civilian Body into the Frame

Greene, in "Drone Vision," describes works by the artist and writer James Bridle that include the body count of civilian deaths in otherwise antiseptic images of violence captured by drone cameras. In *Dronestagram: The Drone's Eye View*, Bridle "collects satellite imagery of drone strike locations and labels them with information from the Bureau of Investigative Journalism about the killing of civilians, combining austere images of desert landscapes and body count figures," adding them to "the smooth flow of Instagram's photo-sharing social media feed" (Greene 239).[5] Bridle argues that "the political and practical possibilities of drone strikes are the consequence of invisible, distancing technologies, and a technologically-disengaged media and society" (*Dronestagram*); in other words, they contribute to the representation and acceptance of a war culture. Social media like Instagram and Twitter connect human users, but Greene uses social media to connect them with military technology and its fatal consequences. Bridle exploits the tension between the supposedly benign media such as Instagram and its implication in images taken from drone strikes. As he says, "technology that was supposed to bring us together is used to obscure and obfuscate," and he links aerial images to body counts as one way to overcome the image of "war without bodies" (*Dronestagram*).

As in Ristelhueber's images, however, there are still no damaged bodies visible in these landscapes. Rather than bodies, Bridle revives the infamous "body count" that falls in and out of favor with the military. In January 2018 in Afghanistan, the U.S. military started publicizing the number of Taliban fighters killed or wounded and posting the results online, but in September the practice was suddenly stopped without explanation. The "body count," though deemed "corrosive" by Defense Secretary Jim Mattis, does at least acknowledge that people are dying, and stopping the practice counters the omission of bodies from coverage of the conflict (Gibbons-Neff). Having no body count makes casualties ungrievable because invisible, an invisibility that visual artists counter with their images.

Photographer Gohar Dashti takes a different approach by posing civilian bodies in combat scenes. In *Today's Life and War*, she juxtaposes ordinary domestic scenes with military imagery. Where Bridle emphasizes the distancing effect of media, Dashti says that the Iran-Iraq War deeply influenced her and her generation, and that "although we may be safe within the walls of our homes, the war continues to reach us through newspapers, television and the Internet" (*Today's Life*). Her images dramatize how the domestic space and war are intertwined, although again they do not show damage, but are arresting for the way in which two supposedly separate realms collide (Figure 8). Like Bridle, she tries to overcome the distancing effect of media coverage of the war. She also sees these as images of survival.

Figure 8. Gohar Dashti, from *Today's Life and War*, 2008. © Gohar Dashti.

Dashti's images convey the military threat to noncombatants supposedly secure in their domestic routines and reinforce the precarity of the civilian body in times of war, such as one in which a couple are eating at a table with a tank looming in the background (Figure 9). The tank looks as if it is about to open fire on the oblivious couple. The disparity between the power of the tank and the domestic scene at the table dramatizes the vulnerability of civilians in times of war and makes visible the invisible threat of military technology such as drones, which kill from a distance. This reintroduction of precarity into images of war contrasts markedly with the military's attempts to create invulnerable soldiers' bodies, which I discuss in the Conclusion.

Where Dashti introduces bodies into the frame of war to evoke precarity, other visual artists use shadows. Øyvind Vågnes, in "Drone Vision: Towards a Critique of the Rhetoric of Precision," foregrounds artworks like Tomas van Houtryve's *Blue Sky Days*, in which civilian bodies are photographed from a drone at times of day when they cast large shadows: "the shadows are both a form of signature and indexical trace, and also appear as a visualization of fragile, precarious life—or even as a kind of prefiguration of death" (15). While Dashti sees the civilians menaced by the tank as an ultimately hopeful image, Houtryve's images are designed to unsettle the viewer and suggest that people carrying out ordinary activities could be the target of a drone strike. The shadows cast by people resemble the outlines of dead bodies drawn at crime scenes, and thus make death an invisible threat for people involved in mundane activities, oblivious to the drone above them. Houtryve's images reinforce a sense of surveillance

Figure 9. Gohar Dashti, from *Today's Life and War*, 2008. © Gohar Dashti.

and precarity as he explores the intersection of "personal privacy, surveillance, and contemporary warfare" (*Blue Sky*). Reports of drones in the United States tend to emphasize their nonthreatening aspects, such as their potentially being used to deliver packages, but Houtryve counters such favorable coverage by showing people going about their lives as potential targets of surveillance and drone strikes. Oddly, one of the few people to voice such possibilities outside of artistic projects was U.S. Senator Ted Cruz, who worried aloud that an American citizen could be the subject of a drone strike "sitting in a coffee shop reading a book" (Nalman). Obviously, in places with active military operations ordinary people could be the target of a drone strike at any time, but Houtryve's images were all collected in the United States, where the civilian population feels itself distanced from and insulated from drone strikes in other countries. As in Dashti's images, such a use of drone vision reinserts the threat to civilian life that is downplayed in the media by its emphasis on the precision of "smart" military technology and its underreporting of deaths of noncombatants.

Similarly, Bridle in his *Drone Shadows* series draws the outline of a drone on city streets to symbolize the invisible and unremarked threat from above.[6] Bridle's approach is close to Ristelhueber's absent bodies in that his "drone shadows" are "not really about what the drone looks like; they're about the absence of the drone in the contemporary discourse" (Drone Center). The proliferation of software like Google Earth that apparently give access to images has led to the illusion that we can "see" what is happening at a distance, but Bridle shows that "for everything that is shown, something is hidden" (*New Dark Age* 36). His

Figure 10. James Bridle, *Drone Shadow 002*, from https://jamesbridle.com/. Courtesy of James Bridle.

"shadows" are like the outline of a corpse but are meant to draw attention to what is left outside the frame of media coverage, which is any meaningful discussion of the human casualties of drone strikes. The size of these huge shadows is important because of the reaction of people when the "the first remark that everyone makes is 'Oh I had no idea they were so big'" (Drone Center). The reaction underscores how little people understand about how military drones operate and their deadly effects, despite media coverage of wars. Bridle underscores the contradiction that people now have a world of information available but are unaware that others are being killed in their name:

> Now, people are being killed by the U.S. and the U.K. in distant lands and we don't even have a picture of where that happens. And yet at the same time, we spent the last 20 years building these network services for digital maps and satellite photography that are supposed to illuminate the whole world. There has been a fundamental disconnect between society and military policy and the techniques of viewing the world which are at our fingertips. (Drone Center)

Bridle's "drone shadows" paradoxically counter the absence of bodies by drawing attention to the invisibility of drones and their threat to civilians. Such images address what Thomas Elsaesser sees as the problem of drones as "vision machines" in that "they generate knowledge that has little to do with human perception or seeing" (Elsaesser 242). Where Ristelhueber reasserts a human presence by casting the landscape as a body, images created by Bridle and Dashti counter the dehumanization of warfare by reinserting the civilian body into the representation of war. Perhaps the most egregious omission from any account of warfare in "war without bodies" is that more civilians are killed than soldiers as a result of violence. According to a UNICEF report on civilian casualties, "civilian fatalities in wartime climbed from 5 per cent at the turn of the century, to 15 per cent during World War I, to 65 per cent by the end of World War II, to more than 90 per cent in the wars of the 1990s" (UNICEF). The bodies of civilians are therefore the most damaged by the mechanization of warfare, which may account for their absence in the media. We are not moving toward a "war without bodies," but a war without soldiers' bodies, in which the "collateral damage" is mainly to the civilian population. While the body count for soldiers may go down with the automation of warfare, war will remain deadly for noncombatants. While the military dreams of invulnerable soldiers thanks to automation, the work of these photographers reminds the viewer of precarity and the vulnerability of the human body.

Conclusion

FUTURE WAR WITHOUT BODIES

When Roger Fenton took photographs of the British military during the Crimean War, the casualties were primarily combatants, although the civilian population of Sevastopol obviously also suffered during the siege. However, such losses were localized to active battle zones, and descriptions and photographs framed the war in terms of the heroism of British forces.[1] The strategies for framing war to make it palatable have changed since Fenton's photographs, in tandem with both military and visual technology, as Virilio has argued in *War and Cinema*. Virilio did not, however, address how military technology has altered the body count.[2] Starting with the Second World War, distant civilian population centers became targets of bombing intent not only on reducing the capacity to wage war, but also on eroding popular support for continuing the conflict. The ratio between combatant and civilian deaths has therefore shifted. Reports of both military and civilian casualties undermine popular support for war, although more research has been carried out on the effect of military casualties on public opinion than on civilian deaths (Johns and Davies 252). The exclusion of body counts and the suppression of images of dead bodies bolster support for a war culture that downplays death, which is the effect of a representation of "war without bodies" that I have analyzed in the preceding pages. A further step toward a "war without bodies" is being researched by the U.S. military. Technology is being developed to reduce combatant casualties through automated warfare that can be waged without soldiers' bodies.

The U.S. military is researching various ways to automate warfare, ranging from completely robotic systems to cyborgs that combine human and artificial intelligence. This research is, of course, highly classified and seems more suited to science fiction, but it fulfills an ambition to negate the influence of random events, which Clausewitz cited as "the fog of greater or lesser uncertainty" in *On War* (26). Lieutenant General Rick Lynch, in an address at a conference, discussed the use of robots in the military and placed them in the context of precarity by saying that "I contend there are things we could do to improve the survivability of our service members" through automation (Pellerin).[3] The military dream is to eradicate precarity by creating either autonomous weapons systems or cyborg supersoldiers.[4] This is a vision of war without soldiers' bodies that represents the culmination of a process that began with the coverage of the Gulf War, where "smart" weapons became the focus as "the drama moved from flesh to weapon" and no bodies were shown in media coverage (Stahl, *Through the Crosshairs* 31). Such a vision of automated warfare is not new; David H. Keller wrote about a fictional war for *Air Wonder Stories* in 1929 in which Japan launches an attack on the United States from Mexico using unmanned aircraft controlled by radio signals, but the plan is thwarted when American radio operators hijack them and guide them harmlessly out to sea ("Bloodless War").

The military's attempt to deny precarity reflects a broader mindset "in which the U.S. subject seeks to produce itself as impermeable, to define itself as protected permanently against incursion and as radically invulnerable to attack" (Butler, *Frames* 47).[5] Patrick Porter's analysis links this military dream of invulnerability to the myth of "absolute security" that undergirded American foreign policy as part of a globalist strategy (84–86). To try to eradicate precarity is to deny the vulnerability of the human body. This emphasis on invulnerability subverts interdependence and severs social responsibility to others who are victims of war. Tim Blackmore sees this as the U.S. military "trying to overcome the various susceptibilities [corporeal, emotional] of the human body" (8) and deny collective vulnerability to death. Pötzsch in turn describes efforts to deny the precarity of the human body as erecting an "epistemological barrier" that creates a "dominant discourse of the self" and implicitly condones the killing of others who are "de-humanized and de-subjectified." ("Borders, Barriers" 78). American military and foreign policy disavow precarity, and in doing so enlist a civilian subjectivity that will support their agenda. Denying the precarity of the body domestically in representations of a "war without bodies" also aligns the American civilian population with its military and foreign policy objectives; war does not seem threatening to their own safety and security, which is, of course, the mindset that Dashti, Bridle, and others are working to subvert.

The increasing automation of war suggests a future in which civilians are collateral damage in battles waged by robots. The *Terminator* film franchise expresses well the fear that machine-driven war will make humans irrelevant to warfare and lead to violence on a mass scale. Science fiction films and novels thus convey warnings about the consequences of "war without bodies." These films register the fear that civilian bodies are increasingly at risk because of the automation of warfare, as shown by the report "Patterns in Conflict: Civilians Are Now the Target" (UNICEF). The bodies of civilians are therefore the most damaged by the mechanization and derealization (Huntemann; Pick; Whitehead and Finnström) of warfare, even as they are erased from contemporary media coverage. In these films a machine is a character, taking on the same role played by drones in the "weapon's eye view" in videos of missile strikes (Stahl, *Through the Crosshairs* 31). Films like *The Terminator* express fears of artificial intelligence and automated warfare, but then assuage such anxieties by having a plucky band of humans triumph in the end. Automated warfare is a dream for the military and a nightmare for civilians because, while soldiers' bodies will be protected, there is no evidence that the increase in civilian casualties recorded by the UNICEF will change.

Armin Krishnan in *Killer Robots* reports that "a declared goal of Pentagon planners is the development of a Terminator-like humanoid military robot, which could fight in human environments such as cities and inside buildings" (75). The military is pursuing "a fantasy of 'indomitability' and invulnerability" (Howell 28) that itself could lead to more conflicts because it discounts the loss of life and damage of war. Krishnan's invocation of the *Terminator* franchise is apt because it exemplifies some of the slippage in the term "robot" that he identifies in his study of the automation of warfare. Science fiction stories like those in the *Terminator* series play upon the different associations of "robots" as "obedient machines following the commands of humans mindlessly" to "machines that can become self-aware and can suddenly decide to follow their own interests" (8); the climactic scene of the first *Terminator* film plays with this juxtaposition by contrasting the killer machine from the future with an automated factory in which "mindless" machines carry out repetitive tasks.[6] There is also an ambiguity in the image of the supersoldier as to whether it is a human with an exoskeleton or other mechanical parts or artificial intelligence in a humanoid shape.[7] Again, the first *Terminator* film captures this ambiguity because Arnold Schwarzenegger as the Terminator appears human at the beginning and is completely mechanical by the end.

In the first *Terminator* film, directed by James Cameron, the robot played by Arnold Schwarzenegger arrives in 1984 Los Angeles completely naked, which allows for several scenes focusing on his nude, muscled torso. Given the connection in American culture between muscles and masculine power (Boyle 47), Schwarzenegger's body is the object of desire but also of fear; Ellexis Boyle notes that "while lingering camera angles invite the audience to admire his uniquely sculpted physique, he is undoubtedly a specter of fear" (46).[8] Kyle Kontour sees the Terminator character as reflecting the "brand new realities of the way in which wars are fought—mediated, virtual, cybernetically and informationally enhanced" and represents "a new definition of military masculinity" (249). Rather than a new definition of masculinity, the Terminator figure represents the fear that the human body will be replaced by machines, thanks to the mechanization of warfare, and that the primary casualties will be human civilians. The first and second *Terminator* films open with scenes of skulls littering the landscape to represent the billions of casualties of a nuclear war started by the artificial intelligence network Skynet. While the filmmakers may not have meant to evoke the association, the aircraft in the opening scenes parallel the current use of drones to carry out violence from afar. A playground that is ruined in the opening of the second film, *Terminator: Judgment Day*, has been destroyed by a nuclear blast, and "judgment day" in the title refers to the death of the billions of civilians as a result of attacks by automated weapons systems gone rogue. In the dystopian future of the film, there are practically no civilians left as these automated weapons systems and human soldiers battle in a ruined Los Angeles.

The *Terminator* films are Gothic tales in the tradition of Mary Shelley's *Frankenstein*, in that the creation of human science rises up to battle its creator; but where the monster was composed of dead bodies in *Frankenstein* (1818), Schwarzenegger's character simulates an organic body but is not human. The plot of the first *Terminator* film follows an arc in which the Terminator's body is gradually revealed to be metallic as his synthetic flesh is burned away. He is first shown performing surgery on his robotic arm, revealing the machinery under his skin, and later loses part of his face to reveal an artificial eye with a red center, which he covers with sunglasses.[9] Visually, the film links the metallic Terminator with the tanks and drone-like aircraft in the opening sequence, where they are battling humans. Given that the Terminator's exterior is made of living tissue that mimics sweating and other bodily functions, the film enacts symbolic violence against the human body in its replacement of flesh by metal, in addition to the billions of humans killed by a nuclear war. The connection between the Schwarzenegger character as flesh and the metallic robot is even more explicit in the second *Terminator* film, which opens with completely metallic robots joining the tanks and aircraft in battle. The film presents a nightmare vision of humans eradicated by military technology as their robot bodies, which have two arms and legs like their human prey, replace their creators.

The *Terminator* films are a visual analogue to Friedrich Kittler's theories discussed in the Introduction. Kittler wrote that "increasingly, data flows once confined to books and later to records and films are disappearing into black holes and boxes that, as artificial intelligences, are bidding us farewell on their way to nameless high commands" (*Gramophone* xxxix). The *Terminator* franchise repeats as horror Kittler's account of the determining power of digital technology when artificial intelligence war machines take over Los Angeles and, presumably, the globe. "High commands" is a translation and may not have had the same ominous, militaristic ring in Kittler's original German as it does in English, where it hovers between computer instructions and control of an army. The overwhelming power of Skynet in *The Terminator* replicates how "the dominant information technologies of the day control all understanding and its illusions," according to Kittler (*Gramophone* xl). Kittler's account erases the human body in making digital technology such an all-consuming force. While the *Termi-*

nator films show an attempt to destroy all human bodies by intelligent machines, they assuage the fears of the viewer by having the human resistance triumph at the end, and they are more optimistic than Kittler's account of the determining power of technology.

Katherine Hayles, by contrast, has argued for the reintroduction of the body and against "an ideology of disembodiment" (192). For Hayles, "human being is embodied being" (283), and she uses this thesis to differentiate human consciousness from that of cybernetic machines (284). As Nicholas Gane says, Hayles asserts "the continued existence of the human body by bringing into view the material practices and interfaces through which bodies and machines meet" (37), while preserving the difference between human and machine embodiment. Rather than "war without bodies," Hayles underlines the difference between human and machine embodiment in a way that would make the Terminator robot impossible. The effort in the second *Terminator* film is to humanize the Schwarzenegger character by making him sensitive to human emotions, and also to become a self-sacrificing savior of humanity as he battles a "new T1000 prototype" that is even more powerful and deadly than he is. In Hayles's terms, this attempts to erase the difference between human and machine embodiment. The fusion does not succeed, because the Terminator at the end of the second film immerses himself in a vat of molten metal, following the fate of the machine that he and his human allies have vanquished.

The *Terminator* films, like the theories that Hayles critiques, are fantasies of unlimited power and disembodied immortality through automation (Hayles 5) that supplant the human body. As with nuclear weapons, they show how the push toward a "war without bodies" is a danger to humans on a mass scale. Works such as those of Ristelhueber, Dashti, Bridle, and others present an important corrective to such fantasies of disembodiment by representing the destruction of bodies and landscapes as a consequence of war. The challenge for artists and writers is to break into a media environment that circulates fantasies of war without material consequences for civilians, and to oppose sanitized coverage that implicitly supports organized state violence, or what Der Derian named the "military-industrial-media-entertainment network." Visons of war without casualties, or waged by sentient machines with metallic bodies, deny the common human bond of precarity and make war thinkable. Artists and writers who reassert the precarity of the human body are an essential counter-narrative to the myth of "war without bodies."

Acknowledgments

The research for this book was enabled by grants and a sabbatical from Brock University, especially a leave in 2019 that gave me time to work in the archives of the British Library and the National Army Museum in London.

My university in general, and my colleagues in the English Department in particular, provide a stimulating and supportive environment for my wide-ranging research agendas. I owe a particular debt of gratitude to my colleague Tim Conley, who, after hearing an early talk on this research project, suggested that *War without Bodies* would be a good title.

Julia Garcia and Jakob Vujovic were both astute and helpful readers of earlier versions of this manuscript.

Sophie Ristelhueber inspired this research project with the image of the crater in figure 6. I teach a course in Victorian literature in which I show Roger Fenton's *Valley of the Shadow of Death* and discuss the charge of the Light Brigade, and I also teach a Lifewriting course that includes memoirs by American soldiers who served in the Iraq War. In that course I show her photographs and had long been fascinated by the possible dialogue between images 137 years apart by her and Fenton. This monograph is the result of several years of teaching and thinking about her photographs. She was incredibly generous in providing me with copies of her photographs and with permission to reproduce them here.

James Bridle was equally generous in providing both the photograph of the drone shadow in figure 10 and permission to use it on the cover. He makes his images available for download at jamesbridle.com in another example of his generosity of spirit.

I was fortunate enough to see Gohar Dashti's photographs in an exhibit at the Institut des Cultures d'Islam in Paris in 2015. The images have stayed with me ever since, and I am grateful that she gave me permission to reproduce them in this monograph.

Finally, my thanks to the anonymous reviewers who gave me such helpful suggestions on how to improve the original versions of this project. Their kind and judicious comments were ideal examples of constructive criticism.

Notes

INTRODUCTION

1. I am not the first person to draw parallels between Fenton and Ristelhueber. David Mellor sees her images as "reminiscent of Fenton" and quotes an earlier 1993 review of an earlier Ristelhueber exhibition in which Ian Walker cited the *Valley of the Shadow of Death* as a comparison (218).

2. While Tom Standage's comparison of the telegraph and the internet erases crucial differences, he is correct that both forms of communication ushered in new social practices.

3. "Freedom" is an ideologically complex term in the United States, invoked in many different political contexts; see also "freedom fries" for how foodstuffs were politicized.

4. There is, of course, a productive ambiguity in "framing" in that it suggests the literal frame around a photograph or painting; Judith Keene exploits this ambiguity in analyzing a painting on the Korean War by Pablo Picasso.

5. See Thomas Gregory's analysis of how Afghan civilian lives were made "ungreivable" through "normative frameworks that determine whose life counts and, effectively, who can appear as human within the public sphere" (333) for an application of Butler's concept of "frames."

6. I am using "interpellation" here in the sense proposed by Louis Althusser as the way in which power calls into being congenial states of consciousness. See "Ideology and Ideological State Apparatuses" (1972).

7. Armitage compares Virilio on "derealization" to Jean Baudrillard's concept of "simulacra (*Virilio*); see, for instance, Baudrillard on the Gulf War as "the delirious spectacle of a war which never happened" (*Gulf* 95). This is also the central premise of Simons and Lucaites's *In/Visible War.*

8. See the essays edited by Paul D'Angelo for examples of frame analysis applied to the news.

9. Sky LaRell Anderson cites a similar "corporeal turn" in the analysis of video games.

10. I am drawing here on Michel Foucault's concept of "governmentality," which links power, the subject, and the state, first elaborated in *Security, Territory, Population.*

11. Mary Favret uses the example of the Napoleonic wars to gauge the "effect of war mediated, brought home through a variety of instruments" (11), and the experiences and emotions she describes in the early nineteenth century also apply to early twenty-first century Texas.

12. See also Reid's application of biopolitics to war in *The Biopolitics of the War on Terror.*

13. Elena Semino has published a prescient warning on the negative effects of war metaphors in the treatment of cancer patients in another example of how violent language produces negative effects.

14. As De Rosa and Peebles note, the United States has not officially declared war since World War II, while at the same time drones and other devices, in addition to the militarization of the police, force "contemporary wars [to] change the ways we understand social space, relationships, and interconnectivity" ("Enduring Operations" 207).

15. See, for instance, the American Civil Liberties Union report on police militarization, *War Comes Home.*

16. See W.J.T. Mitchell, *Cloning Terror* for an extended use of biological metaphors for the "war on terror."

17. This is a euphemistic way of describing an imbalance between a major military power like the United States and a loosely organized insurgency with far fewer resources.

18. See Armitage on Virilio and substitution (9).

CHAPTER 1 — SACRIFICIAL BODIES

1. See, for instance, Monica Cohen, *Professional Domesticity in the Victorian Novel* (1998), Elizabeth Langland, *Nobody's Angels* (1995), and Simon Morgan, *A Victorian Woman's Place* (2007).

2. The most famous example of such a household conduct manual is Sarah Stickney Ellis's *The Women of England: Their Social Duties and Domestic Habits* (1839).

3. Elllis's *The Women of England* extols the domestic habits of English women in contrast to those of France, for example, making housekeeping a source of national pride.

4. J. A. Mangan documents the early inculcation of "militaristic masculinity" in boys in public schools, especially through sport ("Muscular" 157).

5. Hichberger notes that "since Waterloo communications had reached new levels of efficiency. The railways, penny post and, above all, cheap newspapers, meant that the war overseas could be followed by a mass audience" (49).

6. Hannavy claims that Fenton was under direct orders from the War Office not to photograph dead bodies, but the records were destroyed in the Second World War (Lalumia 116), and there is no direct evidence for this assertion. Such a prohibition may have been unnecessary given that Fenton had the "right" class connections and attitudes, unlike Russell, who was Irish and from a lower-class background, more sympathetic to the common soldier than the officer class, and was viewed with some hostility by the upper echelons of the British military, in contrast to Fenton, who dined with them on a regular basis.

7. Felice Beato, by contrast, may have taken the first photographs of dead bodies after a battle, sometimes staging them for effect (see Harris).

8. Jennifer Green-Lewis also notes this contrast between Fenton's diaries and his photographs (107).

9. See http://en.wikipedia.org/wiki/File:Cornet_Wilkin_11th_Hussars.jpg.

10. Karl Baptist von Szathmari was actually the first war photographer (Gernsheim and Gernsheim 10).

11. The photographer had only ten to fifteen minutes in which to develop the image before the plate dried, which is why Fenton traveled with a portable darkroom on a wagon that was always close at hand.

12. See Robert Hirsch (43) and Mary Warner Marien (67) on funerary portraits. Elizabeth Heyert notes that "public demand for funeral portraits, taken after death, grew enormously in the first three decades after the introduction of the photograph," showing that the Victorian public were not averse to photographs of the dead (46).

13. The deceased were often dressed and posed to look as if still alive, which would not have been the case with battlefield corpses (Hilliker 250).

14. Errol Morris carried out extensive forensic modeling to try to determine whether Fenton placed the cannonballs himself and concluded that it was impossible to determine conclusively whether he did so (3–5).

15. Pierre Nora used the idea of "sites of memory" in his three-volume *Les Lieux de Mémoire*, in which he posited a rupture in French culture between "memory" and "history." In Nora's formulation of "sites of memory," Fenton's photograph is part of "history" rather than "memory," and like archives, museums, and monuments represents a self-conscious historicity (13).

16. Natalie Houston has argued that the Victorian sonnet in particular employs "description and memorialization," but Crimean War poetry in general can also be seen as fulfilling these functions (353).

17. *Charge of the Light Brigade* (1877), https://artuk.org/discover/artworks/the-charge-of-the-light-brigade-42648.

18. I am discussing here the public perception of the Victorian soldier. As Furneaux has amply documented, in private correspondence and in memorial books Victorian soldiers could display emotions that were excluded by the stereotype of the "stiff upper lip" for men (16). However, in published images and texts by civilians about the Crimean War, the emphasis is on soldiers as warriors willing to sacrifice themselves in battle.

19. See Kestner Chapter 5, "The Valiant Soldier" (189–234). See also James Eli Adams on heroism and sacrifice in Tennyson's poem (159).

20. The essay "Living with Nikki," discussed in Chapter 4, updates this by using a rifle instead of a saber.

21. Joseph Bristow has labeled war "the stage on which the aristocracy may emblematize the pageantry of its ancestral birthrights," using a metaphor of performance (131). The British military already had a long history of war as spectacle before the Crimean War (see Myerly).

22. See, for example in Waddington, "The flash of our sabres gleams" (40); "And their sabres' gleam in the bright sun-beam" (59); "Their sabres flashing bright and high" (79); "Brightly gleam six hundred sabres" (102); "In air five hundred sabres flash!" (126).

23. In the original French, the "*vision radicale du sacrifice de soi à la guerre, qui aboutit à valoriser la 'mort certaine'*" (63).

24. "Patriotic death in battle was the finest masculine moral virtue" (Mangan and McKenzie 1087).

25. Similarly, Furneaux says the war was viewed "as a righteous crusade against a despotic power . . . and as an opportunity to spread Britain's preferred broadly liberal political model and to secure its military dominance in Europe" (4).

26. Robert McGregor traces the rhetoric of British "freedom" as "the core of a superior masculine culture" back to the confrontation with France and the French Revolution (151–152). "Freedom" continues to be a word with immense ideological weight in the Unites States (see the Introduction).

27. The Gilbert and Sullivan comic opera also plays with situations where characters have competing duties, so that they find themselves caught between imperatives. The word is repeated so many times that it gradually becomes nonsensical.

28. My thanks to Peter O'Neill for suggesting this reference to me.

29. "It's magnificent, but it's not war, it's madness" (my translation).

30. "I am old, I have seen battles, but this is too much" (my translation).

31. Clifford is here a "military man of feeling," like those studied by Furneaux, although he does not appear in her text.

CHAPTER 2 — THE SOLDIER'S BODY AND SITES OF MOURNING

1. Statues of Florence Nightingale and Sidney Herbert were added in front of the memorial in 1915 in a belated recognition of their roles in the Crimean War.

2. Prefiguring this use of the prone male body by Lady Butler, John Everett Millais depicted a soldier returned from the Crimean War in *Peace Concluded, 1856*. A wounded British officer reads about the end of the war, but there is no joy in the painting, mirroring the ambiguity of the result of the conflict. Michael Hanchard reads the image as "man as patient" (506) rather than as the upright, heroic soldier in Fenton's photographs.

3. As Matus notes, nationalism played a role in that these strictures applied only to British soldiers; accounts of the effect of war on French soldiers by British commentators noted the psychological effects of war trauma; since "valor and heroism" were not at stake, the mental trauma could be recognized (51).

4. See Benedict Anderson's *Imagined Communities* (1991) on the use of the press and other media to create a sense of national identity. See also Olive Anderson (70–93).

5. Luckhurst notes that the German (*Granatschock* or *Granatkontusion*) and French (*vent du projectile*) terms also linked bombardment and trauma (51).

CHAPTER 3 — WAR GAMES

1. Gilles Deleuze and Félix Guattari make chess and Go representatives of different approaches to war, the "polis" and "nomas," respectively, in analyzing the "war machine" (4–6).

2. Cecil Eby sees Wells as participating in the "martial spirit" of British popular literature, and *Little Wars* certainly shows this tendency.

3. Originally restricted to members of the Zoological Society, as the century progressed it shifted to "mass consumption" by opening to the public rather than scientific analysis (Jones 2).

4. *Kriegsspiel* was invented for the Prussian military by George Leopold von Reisswitz in 1812 and refined by him in 1824. It aimed to give a realistic simulation of battle tactics as part of military training.

5. Earlier in the century, the Brontës had used toy soldiers as the basis for imaginary warfare, but they were part of a wider fictional universe that was incorporated into their writing (see Butcher).

6. In Wells's *The New Machiavelli* (1910), the main character, Dick Remington, and his friend Britten play miniature war games as Wells's real-life interest found its way into his fiction.

7. Chess was one of the first correspondence games that was played by sending turns to each other in the mail. When two players owned the same Avalon Hill map and game pieces, they could follow the same procedure as chess. With the advent of the internet, play by email (PBEM) games replaced mailing turns. Such games are also known as "turn-based" because only one player moves pieces at a time, as opposed to online real-time games where movement is simultaneous.

8. In a more recent example, the Japanese anime artist Hayao Miyazaki "loathes war and fiercely protests against it," but also "creates manga and films that celebrate the glories of military technology" (Napier 257).

9. See Sara Brady's "War the Video Game" for an extended personal meditation on the conflation of war and games.

10. Digitized versions of the original paper Avalon Hill maps and counters are available on the VASSAL game site for free download. http://www.vassalengine.org/.

11. *Space Invaders*, for instance, created by Tomohiro Nishikado in 1978, was a widely influential video game that depended on the skill and speed of the operator in shooting down "aliens." Nishikado credited *The War of the Worlds* as one of his influences (Puc).

12. There has been much debate on whether on-screen violence increases off-screen violence among players. A meta analysis of twenty-four studies on the association reached the conclusion that "playing violent video games is associated with greater levels of overt physical aggression over time" (Prescott 9887).

13. https://worldofwarcraft.com.

14. I am discussing here what Ian Bogost in *Persuasive Games* has termed the "procedural rhetoric" of video games. See also David Demers; Jasper Juul.

15. This text comes from the DVD boxed version of the game, which is no longer available. The text is available on the websites of various game retailers through an online search.

16. See, for instance, Kneer; Pötzsch; Prescott.

17. Payne sees this interpellation as part of "armaments culture" that recruits civilians into consumption and production of a military entertainment complex (13).

18. In "The Repressive Hypothesis," part of his four-volume *The History of Sexuality*, Foucault discusses how power incites discourse.

19. See also the introduction to Beth Kolko, Lisa Nakamura, and Gilbert Rodman's *Race in Cyberspace*, and Nakamura's analysis of how real-world prejudice informs the mechanics of online fantasy games in "Don't Hate the Player, Hate the Game."

CHAPTER 4 — TRAUMA AND THE SOLDIER'S BODY

1. For an analysis of the media's self-censorship from the Persian Gulf war onward, see Susan Jeffords and Lauren Rabinowitz, *Seeing through the Media* (1994).

2. See, for instance, Amanda Gardner.

3. For the clinical definition of PTSD, see the American Psychiatric Association, *Diagnostic and Statistical Manual of Mental Disorders*, 5th ed. (271–280). For a summary of the symptoms of PTSD, see the fact sheet available at http://www.dsm5.org. PTSD is a controversial diagnosis when applied to the military, because some see it as underdiagnosed and others as overdiagnosed. See Fisher on the political aspects of PTSD and its "prevalence."

4. Not surprisingly, the military takes a teleological approach to PTSD, focusing on the most efficient methods of treatment to overcome the persistent and involuntary return of memory. See, for instance, Erdtmann; Kip; Lee.

5. Pederson engages briefly with the narrative of a veteran in his perceptive analysis but moves on quickly to discuss Hiroshima (342–344).

6. The long, complex debate about the status of victims and viewers of trauma in terms of "witnessing" is summarized in their Introduction by Kilby and Rowland (1–5). The common thread between my discussion of trauma and "witnessing" is how veterans "cope with horror and violence" (5).

7. The report is available online at http://sapr.mil/index.php/reports.

8. Rossiter and Smith refer to sexual assault as "invisible wounds," underscoring the correlation of physical and mental injury. The documentary *The Invisible War*, directed by Kirby Dick, uses personal narratives by servicewomen who have suffered sexual assault to draw attention to the problem.

9. The acceptance of women into active combat units erased one of the major distinctions based on gender in the military.

10. MOS stands for Military Occupational Specialty code, a nine-character code used in the U.S. Army and Marines to identify a specific skill.

11. See, for instance, memoirs by Buzzell and Hartley.

12. As Rossiter and Smith note, 30 percent of female veterans have experienced sexual assault, and a staggering 80 percent have experienced sexual harassment while in the military (366).

13. Williams may have read Irigaray and other feminist theorists, because she took women's studies courses as a student (31).

14. Hartley makes the same point about the hypersexualized environment, using it as an argument against women in the military, rather than suggesting that men might alter their behavior (129).

15. For a detailed analysis of the iconography of the photographs, see Dora Apel's *War Culture and the Contest of Images* (103–111).

16. Colby Buzzell, in *My War*, espouses a different approach, repressing memory altogether. As he says, "I've put the events of that day in a shoebox, and I've never opened it since" (260).

17. The "Warrior Writers" project can be found at https://www.warriorwriters.org.

18. The book edited by Carroll was released in 2006, and the video directed by Richard E. Robbins first aired in 2007.

19. While most texts by Iraq War veterans express guilt, Marcus Luttrell, in *Lone Survivor*, advocates changing the rules of engagement so that U.S. soldiers could shoot on sight (191–193). However, when faced with the decision whether to kill civilians or release them, he and the others in his unit voted to let them go, even though this endangered their lives (234–237).

CHAPTER 5 — SOPHIE RISTELHUEBER

1. Caren Kaplan notes that the aerial view of the landscape in particular is "usually associated with utilitarian state, military, or municipal projects (reconnaissance, surveying, cartography, urban planning)" (4).

2. Clearly this is not true. See, for instance, "US Drone Strike Victims in Pakistan Plan Legal Action," https://www.bbc.com/news/av/world-south-asia-15532916.

3. The *Daily Telegraph*, for instance, proclaimed that "time itself is telegraphed out of existence" (Standage 40).

4. Rachel Carson's *Silent Spring* (1962) marked the beginning of the contemporary ecological movement, although awareness of the effects of industrialization and urbanization on the landscape dates back to the nineteenth century.

5. Thomas Stubblefield in his overview of drone-related art projects addresses Bridle along with others trying to disrupt the frame of drone vision representation (*Drone Art*).

6. Bridle's images can be found at www.jamesbridle.com.

CONCLUSION

1. Leo Tolstoy in *The Sebastopol Sketches* (1855–1856) documented the suffering of the citizens of Sevastopol as well as fictionalized accounts of Russian infantry fighting and dying in battle against the French besiegers.

2. The casualty rate is not the same for all soldiers' bodies. Douglas Kriner and Francis Shen analyze the disproportionate toll on some social groups in terms of class and ethnicity in *The Casualty Gap*.

3. While it may seem odd to apply "precarity" to the military, because it is usually used to refer to vulnerable populations, Nancy Ettlinger in "Precarity Unbound" argues that it should be defined as "physical and nonphysical violence that invokes fear" (321). Soldiers as "implicated subjects" are both bearers of and subject to violence, and thus experience precarity in Ettlinger's terms (see chapter 4).

4. The effort to overcome chance or contingency dates back, according to Engberg-Pedersen's essay, to the eighth century B.C.E., when the Assyrian Emperor Sargon II consulted an astrologer to predict the outcome of a projected campaign ("Wallenstein's" 231). Carolyn Marvin cites the hope in the late nineteenth century that mechanical weapons would end war forever (147).

5. For an extended discussion of Butler's use of vulnerability and the body as a political critique, see Julian Reid's "The Vulnerable Subject."

6. Krishnan gives a history of such "robots" going back to antiquity (14–20).

7. See, for instance, "U.S. Military Plans Cyborg Soldiers with New DARPA Project" (Cuthbertson).

8. Anne Allison sees the use of Schwarzenegger's body as "signaling an unease with the technologization of today's world" (250). I would argue that his body functions as a signifier of male power shifting from the corporeal to the metallic. There are, after all, no female Terminators, and the metallic bodies are clearly male.

9. The single red eye recalls the renegade computer HAL 9000 in Stanley Kubrick's film *2001: A Space Odyssey* (1968).

Works Cited

Abousnnouga, Gill, and David Machin. *The Language of War Monuments*. Bloomsbury, 2013.

Adams, James Eli. "Crimea and the Forms of Heroism." In *A History of Victorian Literature*. James Wiley & Sons, 2009, pp. 156–164.

Adelman, Rebecca A. *Figuring Violence: Affective Investments in Perpetual War*. Fordham University Press, 2019.

Adelman, Rebecca A., and Wendy Kozol. "Unremarkable Suffering: Banality, Spectatorship and War's In/visibilities." In *In/Visible War: The Culture of War in Twenty-First-Century America*, edited by John Simons and John Louis Lucaites. Rutgers University Press, 2017, pp. 89–107.

Ahmad, Muhammad Idrees. "The Magical Realism of Body Counts: How Media Credulity and Flawed Statistics Sustain a Controversial Policy." *Journalism*, vol. 17 no. 1, 2016, pp. 18–43.

Allison, Anne. "Cyborg Violence: Bursting Borders and Bodies with Queer Machines." *Cultural Anthropology*, vol. 16, no. 2, May 2001, pp. 237–265.

Althusser, Louis. "Ideology and Ideological State Apparatuses." In *Lenin and Philosophy and Other Essays*. Monthly Review Press, 1971. https://www.marxists.org/reference/archive/althusser/1970/ideology.htm.

Amato, Sarah. *Beastly Possession: Animals in Victorian Consumer Culture*. University of Toronto Press, 2015.

American Civil Liberties Union. *War Comes Home*. June 2014. https://www.aclu.org/report/war-comes-home-excessive-militarization-american-police.

American Psychiatric Association. *Diagnostic and Statistical Manual of Mental Disorders*, 5th ed. American Psychiatric Publishing, 2013.

Anderson, Benedict. *Imagined Communities: Reflections on the Origin and Spread of Nationalism*, rev. ed. Verso, 1991.

Anderson, Olive. *A Liberal State at War: English Politics and Economics during the Crimean War*. St. Martin's Press, 1967.

Anderson, Sky LaRell. "The Corporeal Turn: At the Intersection of Rhetoric, Bodies, and Video Games." *Review of Communication*, vol. 17, no. 1, pp. 18–36. http://dx.doi.org/10.1080/15358593.2016.1260762.

Apel, Dora. *War Culture and the Contest of Images*. Rutgers University Press, 2012.

Armitage, Paul. "Paul Virilio: An Introduction." *Theory, Culture & Society*, vol. 16, nos. 5–6, 1999, pp.1–23.

Baldwin, Gordon, Malcolm Daniel, and Sarah Greenough. *All the Mighty World: The Photographs of Roger Fenton, 1852–60*. Metropolitan Museum, 2004.

Barczewski, Stephanie L. *Heroic Failure and the British*. Yale University Press, 2016.

Barker, Thomas Jones. "Charge of the Light Brigade." 1877. https://artuk.org/discover/artworks/the-charge-of-the-light-brigade-42648.

Baudrillard, Jean. "The Gulf War Did Not Take Place." In *Hollywood and War: The Film Reader*, edited by J. David Slocum. Routledge, 2006, pp. 303–314.

BBC News. "French Fries Back on House Menu." 2 August 2006. http://news.bbc.co.uk/2/hi/americas/5240572.stm.

Beck, Peter J. *The War of the Worlds: From H. G. Wells to Orson Welles, Jeff Wayne, Steven Spielberg and Beyond*. Bloomsbury, 2016.

Beidler, Philip D. *Beautiful War: Studies in a Dreadful Fascination*. University of Alabama Press, 2016.

Bektas, Yakup. "The Crimean War as a Technological Enterprise." *Notes and Records: The Royal Society Journal of the History of Science*, vol. 71, no. 3, September 2017. https://royalsocietypublishing.org/doi/10.1098/rsnr.2016.0007.

Berman, Nina. "War Is Fun." In *In/Visible War: The Culture of War in Twenty-First-Century America*, edited by John Simons and John Louis Lucaites. Rutgers University Press, 2017, pp. 111–124.

Berridge, L. A. "Off the Chart: The Crimean War in British Public Consciousness." *19: Interdisciplinary Studies in the Long Nineteenth Century*, vol. 20, 2015. http://doi.org/10.16995/ntn.726.

Blackmore, Tim. *War X: Human Extensions in Battlespace*. University of Toronto Press, 2005.

Blitzer, Wolf. "Interview with Condoleezza Rice." *CNN.com*, 8 September 2002. http://transcripts.cnn.com/TRANSCRIPTS/0209/08/le.00.html.

Bogost, Ian. *Persuasive Games: The Expressive Power of Videogames*. MIT Press, 2007.

Borger, Julian. "War Critics Spineless, Says Cheney." *The Guardian*, 18 November 2005. https://www.theguardian.com/world/2005/nov/18/usa.iraq.

Boyle, Ellexis. "The Intertextual Terminator: The Role of Film in Branding 'Arnold Schwarzenegger.'" *Journal of Communication Inquiry*, vol. 34, no. 1, pp. 42–60.

Brady, Sara. "War the Video Game," in *Performance, Politics, and the War on Terror*. Palgrave Macmillan, 2012, pp. 65–102.

Braga, Fernando, and Maria Braga. "Food, Water & Revolution." In *Warrior Writers: Re-Making Sense*, edited by Lovella Calica. Iraq Veterans against the War, 2008, pp. 122–125.

Bridle, James. *Drone Shadows*. 2012. https://jamesbridle.com/works/drone-shadow-002.

———. *Dronestagram: The Drone's Eye View*. 2012. https://jamesbridle.com/works/dronestagram.

———. *New Dark Age: Technology and the End of the Future*. Verso, 2018.

Bristow, Joseph. "Nation, Class and Gender: Tennyson's *Maud* and War." In *Tennyson*, edited by Rebecca Stott. Longman, 1996, pp. 127–147.

Brock, Rita Nakashima, and Gabriela Lettini. *Soul Repair: Recovering from Moral Injury after War*. Beacon Press, 2012.

Brown, Kenneth D. "Modelling for War? Toy Soldiers in Late Victorian and Edwardian Britain." *Journal of Social History*, vol. 24, no. 2, winter 1990, pp. 237–254.

Butcher, Emma. "War Trauma and Alcoholism in the Early Writings of Charlotte and Branwell Brontë." *Journal of Victorian Culture*, vol. 22, no. 4, 2017, pp. 465–481.

Butler, Elizabeth. *An Autobiography: With Illustrations from Sketches by the Author.* Constable & Co., 1922.

Butler, Judith. *Frames of War: When Is Life Grievable?* Verso, 2009.

———. *Precarious Life: The Powers of Mourning and Violence.* Verso, 2006.

Buzzell, Colby. *My War: Killing Time in Iraq.* Penguin, 2005.

Calica, Lovella, editor. *Warrior Writers: Re-Making Sense.* Iraq Veterans against the War, 2008.

Cameron, Drew. "Living without Nikki." In *Warrior Writers: Re-Making Sense*, edited by Lovella Calica. Iraq Veterans against the War, 2008, p. 80.

———. "You Are Not My Enemy." In *Not My Enemy by Warrior Writers*, edited by Lovella Calica. Brooklyn Artists Alliance, n.d., p. 3.

Cameron, James, director. *The Terminator.* Orion Pictures, 1984.

Camlot, Jason. "Alfred, Lord Tennyson, 'The Charge of the Light Brigade' (1854)." *Victorian Review*, vol. 35, no. 1, April 2009, pp. 27–32.

Campbell, David. "How Photojournalism Has Framed the War in Afghanistan." In *In/Visible War: The Culture of War in Twenty-First-Century America*, edited by John Simons and John Louis Lucaites. Rutgers University Press, 2017, pp. 27–47.

Carroll, Andrew, editor. *Operation Homecoming: Iraq, Afghanistan, and the Home Front in the Words of U.S. Troops and Their Families.* Random House, 2006.

Carruthers, Susan L. "Missing in Authenticity? Media War in the Digital Age." In *War and the Media: Reportage and Propaganda, 1900–2003*, edited by Mark Connelly and David Welch. I. B. Tauris, 2005.

Carson, Rachel. *Silent Spring.* Houghton Mifflin, 1962.

Caruth, Cathy. "Introduction: Trauma and Experience." In *Trauma: Explorations in Memory*, edited by Cathy Caruth. Johns Hopkins University Press, 1995, pp. 3–12.

———. *Unclaimed Experience: Trauma, Narrative, and History.* Johns Hopkins University Press, 1996.

Cavarero, Adriana. *Horrorism: Naming Contemporary Violence.* Translated by William McCuaig. Columbia University Press, 2009.

Chouliaraki, Lilie. *The Spectatorship of Suffering.* Sage, 2006.

Choy, Christopher Yi-Han. "British War-Gaming, 1870–1914." M.A. thesis, King's College Cambridge, 2013. https://www.academia.edu/9691144/British_Wargaming_1870_1914.

Clausewtiz, Karl Von. *On War.* Translated by J. J. Graham. Digireads, 2018.

Clifford, Henry. *Henry Clifford: His Letters and Sketches from the Crimea.* E. P. Dutton, 1956.

CNN. "'Shock and Awe' Campaign Underway in Iraq." 22 March 2003. http://edition.cnn.com/2003/fyi/news/03/22/iraq.war/.

Cohen, Monica. *Professional Domesticity in the Victorian Novel: Women, Work and Home.* Cambridge University Press, 1998.

Craig, Kathleen. "Dead in Iraq: It's No Game." *Wired*, 6 June 2006. http://www.wired.com/gaming/gamingreviews/news/2006/06/71052.

Creative Assembly, The. "High Elves." In *Total Warhead Wiki*. Gamepedia, 12 June 2020. https://totalwarwarhammer.gamepedia.com/High_Elves#Background.

Cuthbertson, Anthony. "U.S. Military Plans Cyborg Soldiers with New DARPA Project." *Newsweek*, 21 January 2016. https://www.newsweek.com/us-military-plans-cyborg-soldiers-new-darpa-project-418128.

Danahay, Martin. "'Valiant Lunatics': Heroism and Insanity in British and Russian Reactions to the Charge of the Light Brigade." In *BRANCH: Britain, Representation and Nineteenth-Century History*, edited by Dino Franco Felluga. June 2018. http://www.branchcollective.org/?ps_articles=martin-danahay-valiant-lunatics-heroism-and-insanity-in-british-and-russian-reactions-to-the-charge-of-the-light-brigade.

D'Angelo, Paul. *Doing News Framing Analysis II: Empirical and Theoretical Perspectives*. Routledge, 2018.

Dashti, Gohar. "Chercher L'erreur." Institut des Cultures d'Islam, Paris, 15 January–19 April 2015. https://www.institut-cultures-islam.org/agenda/categorie-evenement/cherchez-lerreur/

———. *Today's Life and War*. http://gohardashti.com/work/.

Dawson, Graham. *Soldier Heroes: British Adventure, Empire and the Imagining of Masculinities*. Routledge, 1994.

Deleuze, Gilles, and Félix Guattari. *Nomadology: The War Machine*. Translated by Brian Massumi. Wormwood Distribution, 2010.

Demause, Lloyd. "War as Sacrificial Ritual." *Journal of Psychohistory*, vol. 35, no. 3, winter 2008, pp. 231–239.

Demers, David. "The Procedural Rhetoric of War: Ideology, Recruitment, and Training in Military Videogames." M.A. thesis, Concordia University, Montreal, Quebec, 2014.

Der Derian, James. *Virtuous War: Mapping the Military-Industrial-Media-Entertainment Network*. Westview Press, 2001.

Dereli, Cynthia. *A War Culture in Action: A Study of the Literature of the Crimean Period*. Peter Lang, 2003.

De Rosa, Aaron, and Stacy Peebles. "Enduring Operations: Narratives of the Contemporary Wars." *MFS: Modern Fiction Studies*, vol. 63, no. 2, summer 2017.

Descartes, René. *Discourse on the Method of Rightly Conducting One's Reason and of Seeking Truth in the Sciences*, part 5. 1637. https://www.bartleby.com/34/1/5.html.

Dick, Kirby, director. *The Invisible War*. The Documentary Group, 2012.

Dooling, Michael Carroll. "The Thin Iron Line." *Naval History*, vol. 18, no. 3, June 2004, p. 36.

Dowden, E. "Tennyson as the Poet of Law." In *Tennyson: The Critical Heritage*, edited by John D. Jump. Routledge, 1967, pp. 322–333.

Drone Center, Bard College. "Interview: James Bridle." 24 January 2017. https://dronecenter.bard.edu/interview-james-bridle.

Duberly, Frances Isabella. *The Two Wars of Mrs. Duberly*. Leonaur, 2009.

Dunn, David Hastings. "Drones: Disembodied Aerial Warfare and the Unarticulated Threat." *International Affairs*, vol. 89, no. 5, 2013, pp. 1237–1246.

Eby, Cecil Degrotte. *The Road to Armageddon: The Martial Spirit in English Popular Literature, 1870–1914*. Duke University Press, 1988.

Eisenman, Stephen. *The Abu Ghraib Effect*. Reaktion Books, 2007.

Ellis, Sarah Stickney. *The Women of England, Their Social Duties, and Domestic Habits*. Fisher, Son & Company, 1838.

Elsaesser, Thomas. "The 'Return' of 3-D: On Some of the Logics and Genealogies of the Image in the Twenty-First Century." *Critical Inquiry*, vol. 39, no. 2, 2013, pp. 217–246.

Engberg-Pedersen, Anders. "Technologies of Experience: Harun Farocki's Serious Games and Military Aesthetics." *Boundary 2*, vol. 44, no. 4, 2017, pp. 155–178.

———. "Wallenstein's Contingency Media." *Romanticism*, vol. 24, no. 3, 2018, pp. 231–244.

Entman, Robert M. "Framing: Toward Clarification of a Fractured Paradigm." *Journal of Communication*, vol. 43, no. 4, 1993, pp. 51–58.

Erdtmann, Frederick. "Treatment for Posttraumatic Stress Disorder in Military and Veteran Populations: Final Assessment." *Military Medicine*, vol. 179, no. 12, December 2014, pp. 1401–1403.

Ettlinger, Nancy. "Precarity Unbound." *Alternatives*, vol. 32, 2017, pp. 319–340.

Eubanks, Charlotte. "Playing at Empire: The Ludic Fantasy of *Sugoroku* in Early Twentieth-Century Japan." *Verge: Studies in Global Asias*, vol. 2, no. 2, pp. 36–57.

"Every Man of Common Modesty Must Feel. . . ." *Times of London*, 12 October 1854, p. 6.

Farocki, Harun. "Serious Games." *Intervalla*, no. 2, 2014. https://www.fus.edu/intervalla/volume-2-trauma-abstraction-and-creativity.

Fassin, Didier, and Richard Rechtman. *The Empire of Trauma: An Inquiry into the Condition of Victimhood*. Princeton University Press, 2009.

Favret, Mary. *War at a Distance: Romanticism and the Making of Modern Wartime*. Princeton University Press, 2009.

Figes, Orlando. *The Crimean War: A History*. Metropolitan Books, 2010.

Filewod, Alan. "Warplay: Spectacle, Performance, and (Dis)Simulation of Combat." In *Bearing Witness: Perspectives on War and Peace from the Arts and Humanities*, edited by Sherill Grace, Patrick Imbert, and Tiffany Johnstone. McGill-Queen's University Press, 2012, pp. 17–27.

Fisher, Michael P. "PTSD in the U.S. Military, and the Politics of Prevalence." *Social Science & Medicine*, no. 115, August 2014, pp. 1–9.

Fleming, Angela Michelli, and John Maxwell Hamilton, editors. *The Crimean War as Seen by Those Who Reported It: William Howard Russell and Others*. Louisiana State University Press, 2009.

Flint, Kate. *The Woman Reader, 1837–1914*. Oxford University Press, 1993.

Ford, Dom. "'eXplore, eXpand, eXploit, eXterminate': Affective Writing of Postcolonial History and Education in *Civilization V*." *Game Studies*, vol. 16, no. 2, 2016. http://gamestudies.org/1602/articles/ford.

Foucault, Michel. *The History of Sexuality*, vol. 1: *An Introduction*. Translated by Robert Hurley. Pantheon Books, 1978.

———. *Security, Territory, Population: Lectures at the Collège de France 1977–1978*. Translated by Graham Burchell. Picador, 2009.

Friedrich, Ernst. *Krieg dem Kriege!* [*War against War!*], 3rd ed. Spokesman Press, 2014. (Originally published 1924.)

Fuller, Linda K. "Victims, Villains, and Victors: Mediated Wartime Images of Women." In *Women, War, and Violence*, edited by Robin M. Chandler, Lihua Wang, and Linda K. Fuller. Palgrave Macmillan, 2010, pp. 59–72.

Furneaux, Holly. *Military Men of Feeling: Emotion, Touch, and Masculinity in the Crimean War*. Oxford University Press, 2016.

———. "Victorian Masculinities, or Military Men of Feeling: Domesticity, Militarism, and Manly Sensibility." In *The Oxford Handbook of Victorian Literary Culture*, edited by Juliet John. Oxford University Press, 2016, pp. 211–232.

Gane, Nicholas. "Radical Post-Humanism: Friedrich Kittler and the Primacy of Technology." *Theory, Culture & Society*, vol. 22, no. 3, 2005, pp. 25–41.

Garago, Jim. "Experts: Males Are Also Victims of Sexual Assault." *U.S. Department of Defense News*, 20 February 2015. https://www.defense.gov/News/Article/Article/604140.

Gardner, Amanda. "Depression, PTSD Plague Many Iraq Vets." *CNN.com*, 7 June 2010. http://www.cnn.com/2010/HEALTH/06/07/iraq.vets.ptsd.

Gernsheim, Helmut, and Alison Gernsheim. *Roger Fenton: Photographer of the Crimean War*. Secker and Warburg, 1954.

Gibbons-Neff, Thomas. "US Retreats on Publicizing Body Count of Militants Killed in Afghanistan." *New York Times*, 20 September 2018. https://www.nytimes.com/2018/09/20/us/politics/military-body-count-afghanistan.html.

Godbey, Emily. "'Terrible Fascination': Civil War Stereographs of the Dead." *History of Photography*, vol. 6, no. 3, 2012, pp. 265–274.

Godfrey, Sima. "La Guerre de Crimeé n'aura pas Lieu." *French Cultural Studies*, vol. 27, no. 1, 2016, pp. 3–19.

Goffman, Erving. *Frame Analysis: An Essay on the Organization of Experience*. Harvard University Press, 1974.

"Grand Military Spectacle: The Heroes of the Crimea Inspecting the Field-Marshals." *Punch*, 3 November 1855.

Green-Lewis, Jennifer. *Framing the Victorians: Photography and the Culture of Realism*. Cornell University Press, 1996.

Greene, Daniel. "Drone Vision." *Surveillance & Society*, vol. 13, no.2, pp. 233–249.

Grehan, John, and Martin Mace. *British Battles of the Crimean Wars 1854–1856*. Pen & Sword Books, 2014.

Gregory, Derek. "From a View to a Kill: Drones and Late Modern War." *Theory, Culture and Society*, vol. 28, nos. 7–8, pp. 188–215.

Gregory, Thomas. "Potential Lives, Impossible Deaths: Afghanistan, Civilian Casualties and the Politics of Intelligibility." *International Feminist Journal of Politics*, vol. 14, no. 3, September 2012, pp. 327–347.

Groth, Helen. "Technological Mediations and the Public Sphere: Roger Fenton's Crimea Exhibition and 'The Charge of the Light Brigade.'" *Victorian Literature and Culture*, vol. 30, no. 2, 2002, pp. 553–570.

Gustafsson, Henrik. "The Cut and the Continuum: Sophie Ristelhueber's Anatomical Atlas." *History of Photography*, vol. 40, no. 1, pp. 67–86.

Gygax, Gary, and Jeff Perren. *Chainmail: Rules for Medieval Miniatures*, 3rd ed. Tactical Studies Rules, 1975.

Hammar, Emil Lundedal, and Jamie Woodcock. "The Political Economy of Wargames: The Production of History and Memory in Military Video Games." In *War Games: Memory, Militarism and the Subject of Play*, edited by P. Hammond and H. Pötzsch. Bloomsbury, 2019, pp. 54–71.

Hanchard, Michael. "'Urgent Private Affairs': Millais's *Peace Concluded, 1856*." *Burlington Magazine*, vol. 133, no. 1061, August 1991, pp. 499–506.

Hannavy, John. *The Camera Goes to War*. Scottish Arts Council, 1974.

Haraldsson, Hrafknell. "Saving You from Socialist Light Bulbs by Saving You from Pesky Facts." *PoliticusUSA*, no. 3, June 2012. https://www.politicususa.com/2012/06/03/saving-socialist-light-bulbs-saving-pesky-facts.html.

Harris, David. *Of Battle and Beauty: Felice Beato's Photographs of China*. Santa Barbara Museum of Art, 1999.
Hartley, Jason Christopher. *Just Another Soldier: A Year on the Ground in Iraq*. Harper, 2006.
Hayles, Katherine. *How We Became Posthuman*. University of Chicago Press, 1999.
Heyert, Elizabeth. *The Glass-House Years: Victorian Portrait Photography 1839–1870*. Allanheld, Osmun & Co., 1979.
Hichberger, Joan W. M. *Images of the Army: The Military in British Art, 1815–1914*. Manchester University Press, 1988.
Higgin, Tanner. "Blackless Fantasy: The Disappearance of Race in Massively Multiplayer Online Role-Playing Games." *Games and Culture*, vol. 4, no. 1, 2009, pp. 3–26.
Hilliker, Laurel. "Letting Go while Holding On: Postmortem Photography as an Aid in the Grieving Process." *Illness, Crisis & Loss*, vol. 14, no. 3, 2006, pp. 245–269.
Hindry, Ann. *Sophie Ristelhueber*. Hazan, 1998.
Hirsch, Robert. *Seizing the Light: A Social History of Photography*. McGraw-Hill Higher Education, 2009.
Ho, Tai-Chun. "Tyrtaeus and the Civilian Poet of the Crimean War." *Journal of Victorian Culture*, vol. 22, no. 4, 2017, pp. 503–520.
Holmqvist, Caroline. "Undoing War: War Ontologies and the Materiality of Drone Warfare." *Millennium: Journal of International Studies*, vol. 41, no. 3, 2013, pp. 535–552.
Hoskins, Andrew, and Ben O'Loughlin. "Arrested War: The Third Phase of Mediatization." *Information, Communication & Society*, vol. 18, no. 11, 2015, pp. 1320–1338.
Houston, Natalie M. "Reading the Victorian Souvenir: Sonnets and Photographs of the Crimean War." *Yale Journal of Criticism*, no. 14, 2001, pp. 353–383.
Houtryve, Tomas Van. *Blue Sky Days*. n.d. https://tomasvh.com/works/blue-sky-days/.
Howard, Matt. "PTSD." In *Warrior Writers: Re-Making Sense*, edited by Lovella Calica. Iraq Veterans against the War, 2008, p. 163.
Howell, Alison. "Resilience, War, and Austerity: The Ethics of Military Human Enhancement and the Politics of Data." *Security Dialogue*, vol. 46, no. 1, 2015, pp. 15–31.
Huntemann, Nina B., and Matthew Thomas Payne. *Joystick Soldiers: The Politics of Play in Military Video Games*. Routledge, 2010.
Ignatieff, Michael. *Virtual War: Kosovo and Beyond*. Viking Penguin, 2000.
Irigaray, Luce. *This Sex Which Is Not One*. Cornell University Press, 1985.
Jaffe, Greg. "War in Iraq Will Be Called 'Operation New Dawn' to Reflect Reduced U.S. Role." *Washington Post*, 19 February 2010. http://www.washingtonpost.com/wp-dyn/content/article/2010/02/18/AR2010021805888.html.
Jeffords, Susan, and Lauren Rabinovitz. *Seeing through the Media: The Persian Gulf War*. Rutgers University Press, 1994.
Johns, Robert, and Graeme A. M. Davies. "Civilian Casualties and Public Support for Military Action: Experimental Evidence." *Journal of Conflict Resolution*, vol. 63, no. 1, 2019, pp. 251–281.
Johnstone, J. B. *Balaclava: A Drama in Three Acts*. National Standard Theater, Bishopsgate, 10 June 1878. Victoria and Albert Museum Theatre and Performance Collection.
Jones, Robert W. "'The Sight of Creatures Strange to Our Clime': London Zoo and the Consumption of the Exotic." *Journal of Victorian Culture*, vol. 2, no. 1, spring 1997, pp. 1–26.

Juul, Jesper. *Half-Real: Video Games between Real Rules and Fictional Worlds.* MIT Press, 2005.

Kahn, Paul W. "Imagining Warfare." *European Journal of International Law*, vol. 24, no. 1, February 2013, pp. 199–226.

Kaplan, Caren. *Aerial Aftermaths: Wartime from Above.* Duke University Press, 2017.

Kaplan, E. Ann. *Trauma Culture: The Politics of Terror and Loss in Media and Literature.* Rutgers University Press, 2005.

Kaye, John William. *Lives of Indian Officers.* 2 vols. W. H. Allen, 1889.

Keene, Judith. "Framing Violence, Framing Victims: Picasso's Forgotten Painting of the Korean War." *Cultural History*, vol. 6, no. 1, 2017, pp. 80–101.

Keller, David H. "The Bloodless War." In *Fighting the Future War*, edited by Frederic Krome. Routledge, 2012, pp, 160–169.

Keller, Ulrich. *The Ultimate Spectacle: A Visual History of the Crimean War.* Gordon and Breach Publishers, 2001.

Kellner, Douglas. "Virilio, War and Technology: Some Critical Reflections." *Theory, Culture & Society*, vol. 16, nos. 5–6, 1 December 1999, pp. 103–125.

Kendrick, Michelle. "Kicking the Vietnam Syndrome." In *Seeing through the Media: The Persian Gulf War*, edited by Susan Jeffords and Lauren Rabinovitz. Rutgers University Press, 1994, pp. 59–76.

Kessler, Glenn. "The Iraq War and WMDs: An Intelligence Failure or White House Spin?" *Washington Post*, 22 March 2019. https://www.washingtonpost.com/politics/2019/03/22/iraq-war-wmds-an-intelligence-failure-or-white-house-spin/?noredirect=on&utm_term=.452b647d86b9.

Kestner, Joseph A. *Masculinities in Victorian Painting.* Scolar Press, 1995.

Kilby, Jane, and Antony Rowland, editors. "Introduction." In *The Future of Testimony: Interdisciplinary Perspective on Witnessing.* Routledge, 2014, pp. 1–16.

King, C. Richard and David J. Leonard. "Wargames as a New Frontier: Securing American Empire in Virtual Space." In *Joystick Soldiers: The Politics of Play in Military Video Games*, edited by Nina B. Huntemann and Matthew Thomas Payne. Routledge, 2010, pp. 91–105.

King, Cynthia, and Paul Martin Lester. "Photographic Coverage during the Persian Gulf and Iraqi Wars in Three U.S. Newspapers." *Journalism & Mass Communication Quarterly*, vol. 82, no. 3, autumn 2005, pp. 623–637.

King, Erika G., and Robert A. Wells. *Framing the Iraq War Endgame: War's Denouement in an Age of Terror.* Palgrave Macmillan, 2009.

Kip, Kevin E., Laney Rosenzweig, Diego F. Hernandez, Amy Shuman, Kelly L. Sullivan, Christopher J. Long, James Taylor, Stephen McGhee, Sue Ann Girling, Trudy Wittenberg, Frances M. Sahebzamani, Cecile A. Lengacher, Rajendra Kadel, and David M. Diamond. "Randomized Controlled Trial of Accelerated Resolution Therapy (ART) for Symptoms of Combat-Related Post-Traumatic Stress Disorder (PTSD)." *Military Medicine*, vol. 178, no. 12, December 2013, pp. 1298–1309.

Kipling, Rudyard. *Kim.* Macmillan, 1966.

———. *Kipling Stories and Poems Every Child Should Know*, book 2. Edited by Mary E. Burt and W. T. Chapin. Houghton Mifflin, 1891.

Kittler, Friedrich A. *Discourse Networks 1800/1900.* Translated by Michael Metteer, with Chris Cullens. Stanford University Press, 1990.

———. *Gramophone, Film, Typewriter.* Translated by Geoffrey Winthrop-Young and Michael Wutz. Stanford University Press, 1999.

Kline, Stephen, Nick Dyer-Witheford, and Greg de Peteur. *Digital Play: The Interaction of Technology, Culture, and Marketing.* McGill-Queen's University Press, 2014.

Klinke, Ian, and Mark Bassin. "Introduction: Lebensraum and Its Discontent." *Journal of Historical Geography*, vol. 61, 2018, pp. 53–58.

Kneer, Julia, Malte Elson, and Florian Knapp. "Fight Fire with Rainbows: The Effects of Displayed Violence, Difficulty, and Performance in Digital Games on Affect, Aggression, and Physiological Arousal." *Computers in Human Behavior*, no. 54, 2016, pp. 142–148.

Knightley, Phillip. "Here Is the Patriotically Censored News." *Index on Censorship*, no. 20, April/May 1991, pp. 4–5.

Kolko, Beth E., Lisa Nakamura, and Gilbert B. Rodman. "Introduction." In *Race in Cyberspace*, edited by Beth E. Kolko, Lisa Nakamura, and Gilbert B. Rodman. Routledge, 2000, pp. 1–13.

Kontour, Kyle. "War, Masculinity, and Gaming in the Military Industrial Complex: A Case Study of *Call of Duty 4: Modern Warfare*." Doctoral dissertation, University of Colorado, 2011.

Kriner, Douglas and Francis Shen. *The Casualty Gap: The Causes and Consequences of American Wartime Inequalities.* Oxford University Press, 2010.

Krishnan, Armin. *Killer Robots: Legality and Ethicality of Autonomous Weapons.* Routledge, 2009.

Lagrange, François. "Les Combattants de la 'Mort Certaine': Les Sens du Sacrifice a L'horizon de la Grande Guerre." *Cultures et Conflits*, no. 64, December 2006, pp. 63–81.

Lalumia, Matthew. *Realism and Politics in Victorian Art of the Crimean War.* University of Michigan Research Press, 1984.

Lamb, Christina. *Our Bodies, Their Battlefields: War through the Lives of Women.* Scribner, 2020.

Lamothe, Dan. "Sexual Assault on Both Men and Women in the Military Is Declining, Pentagon Survey Finds." *Washington Post*, 1 May 2017. https://www.washingtonpost.com/news/checkpoint/wp/2017/05/01/sexual-assault-on-both-men-and-women-in-the-military-is-declining-pentagon-survey-finds/?utm_term=.d7e266bafc35.

Laneyrie-Dagen, Nadeije. "Examples from Painting." In *A History of Virility*, edited by Alain Corbin, Jean-Jacques Courtine, and Georges Vigarello. Columbia University Press, 2011, pp. 114–145.

Langland, Elizabeth. *Nobody's Angels: Middle-Class Women and Domestic Ideology in Victorian Culture.* Cornell University Press, 1995.

Lee, D. J. "Advances and Controversies in Military Posttraumatic Stress Disorder Screening." *Current Psychiatry Reports*, vol. 16, no. 9, September 2014, pp. 1–6.

Leech, John. "Enthusiasm of Paterfamilias on Reading the Report of the Grand Charge of British Cavalry on the 25th." *Punch* 27, 1854, p. 213.

Leonard, David. "Unsettling the Military Entertainment Complex: Video Games and a Pedagogy of Peace." *Simile*, vol. 4, no. 4, November 2004, pp. 1–8.

Lepianka, Nigel, and Deanna Stover, editors. "Introduction." In *Little Wars*, by H. G. Wells. *Scholarly Editing*, vol. 38, 2017. http://scholarlyediting.org/2017/editions/littlewars/fulltext.html#page_info.

Lewis, Jack. "Road Work." In *Operation Homecoming: Iraq, Afghanistan, and the Home Front in the Words of U.S. Troops and Their Families*, edited by Andrew Carroll. Random House, 2006, pp. 123–126.

Lilley, Terry G., Joel Best, Benigno E. Aguirre, and Kathleen S. Lowney. "Magnetic Imagery: War-Related Ribbons as Collective Display." *Sociological Inquiry*, vol. 80, no. 2, May 2010, pp. 313–321.

Loughran, Tracey. *Shell-Shock and Medical Culture in First World War Britain*. Cambridge University Press, 2017.

Luckhurst, Roger. *The Trauma Question*. Routledge, 2008.

Luttrell, Marcus, with Patrick Robinson. *Lone Survivor*. Little, Brown and Company, 2007.

Lutwak, Nancy. "Military Sexual Trauma Increases Risk of Post-Traumatic Stress Disorder and Depression Thereby Amplifying the Possibility of Suicidal Ideation and Cardiovascular Disease." *Military Medicine*, vol. 178, no. 4, April 2013, pp. 359–336.

Maguen, Shira, and Brett Litz. "Moral Injury in the Context of War." U.S. Department of Veterans Affairs, National Center for PTSD, 23 February 2016. https://www.ptsd.va.gov/professional/co-occurring/moral_injury_at_war.asp.

Maguire, James Rochefort. *Cecil Rhodes; A Biography and Appreciation by Imperialist, with Personal Rèminiscences by Dr. Jameson*. Macmillan, 1897.

Mangan, J. A. "'Muscular, Militaristic and Manly': The Middle-Class Hero as Moral Messenger." *International Journal of the History of Sport*, vol. 27, nos. 1–2, January–February 2010, pp. 150–168.

Mangan, J. A., and C. McKenzie. "'Duty unto Death'—the Sacrificial Warrior: English Middle-Class Masculinity and Militarism in the Age of the New Imperialism." *International Journal of the History of Sport*, vol. 25, no. 9, August 2008, pp. 1080–1105.

Mantello, Peter. "Military Shooter Video Games and the Ontopolitics of Derivative Wars and Arms Culture." *American Journal of Economics and Sociology*, vol. 76, no. 2, March 2017, pp. 483–521.

Marien, Mary Warner. *Photography: A Cultural History*. Harry N. Abrams, 2002.

Markovits, Stefanie. *The Crimean War in the British Imagination*. Cambridge University Press, 2009.

Marston, John Westland. *The Death-Ride: A Tale of the Light-Brigade*. C. Mitchell, 1855.

Martineau, Harriet. *Harriet Martineau's Autobiography*, vol. 1. James R. Osgood and Co., 1877. https://oll.libertyfund.org/titles/martineau-harriet-martineaus-autobiography-vol-1.

Marvin, Carolyn. *When Old Technologies Were New: Thinking about Electric Communication in the Late Nineteenth Century*. Oxford University Press, 1988.

Matus, Jill L. *Shock, Memory and the Unconscious in Victorian Fiction*. Cambridge University Press, 2009.

McDonald, Lynn, editor. *The Collected Works of Florence Nightingale*, vol. 14. Wilfrid Laurier University Press, 2010.

McGann, Jerome. *The Beauty of Inflections: Literary Investigations in Historical Method and Theory*. Clarendon Press, 1985.

McGhee, Richard D. *Marriage, Duty and Desire in Victorian Poetry and Drama*. University Press of Kansas, 1980.

McGregor, Robert. "The Popular Press and the Creation of Military Masculinities in Georgian Britain." In *Military Masculinities: Identity and the State*, edited by Paul R. Higate. Praeger, 2003.

McLuhan, Marshall. *Understanding Media: The Extension of Man*. Routledge, 2004. (Originally published 1964.)

McNally, Richard J. "The Expanding Empire of Psychopathology: The Case of PTSD." *Psychological Inquiry*, no. 27, 2016, pp. 46–49.

———. *Remembering Trauma*. Belknap Press, 2003.

McSorley, Kevin. *War and the Body: Militarisation, Practice and Experience*. Routledge, 2013.

Mellor, David. "Rents in the Fabric of Reality: Contexts for Sophie Ristelhueber." In *Sophie Ristelhueber: Operations*. Thames & Hudson, 2009, pp. 213–228.

Mill, John Stuart. *The Subjection of Women*. Project Gutenberg, 2008. http://www.gutenberg.org/files/27083/27083-0.txt.

Mitchell, W.J.T. *Cloning Terror: The War of Images, 9/11 to the Present*. Chicago University Press, 2011.

Monegal, Antonio. "Picturing Absence: Photography in the Aftermath." *Journal of War & Culture Studies*, vol. 9, no. 3, pp. 252–270.

Morgan, Simon. *A Victorian Woman's Place: Public Culture in the Nineteenth Century London*. Tauris Academic Studies, 2007.

Morris, Errol. *Believing Is Seeing: Observations on the Mysteries of Photography*. Penguin Press, 2011.

Mukherjee, Souvik. "The Playing Fields of Empire: Empire and Spatiality in Video Games." *Journal of Gaming & Virtual Worlds*, vol. 7, no. 3, pp. 299–315.

———. "Playing Subaltern: Video Games and Postcolonialism." *Games and Culture*, vol. 13, no. 5, 2018, pp. 504–520.

Mukherjee, Souvik and Hammar Emile L., "Introduction to the Special Issue on Postcolonial Perspectives in Game Studies." *Open Library of Humanities*, vol. 4, no. 22, 2018, p. 33.

Müller, Simone M. *Wiring the World: The Social and Cultural Creation of Global Telegraph Networks*. Columbia University Press, 2016.

Myerly, Scott Hughes. *British Military Spectacle: From the Napoleonic Wars through the Crimea*. Harvard University Press, 1996.

Nakamura, Lisa. "Don't Hate the Player, Hate the Game: The Racialization of Labor in *World of Warcraft*." *Critical Studies in Media Communication*, vol. 26, no. 2, 2009, pp. 128–144.

Nalman, Robert. "What Rand Paul and Ted Cruz Exposed about the Drone Strikes." *Huffington Post*, 7 March 2013. https://www.huffingtonpost.com/robert-naiman/what-rand-paul-ted-cruz-e_b_2828517.html.

Napier, Susan Jolliffe. *Miyazakiworld: A Life in Art*. Yale University Press, 2018.

Nightingale, Florence. *Cassandra: An Essay*. Feminist Press, 1979.

Nixon, Rob. *Slow Violence and the Environmentalism of the Poor*. Harvard University Press, 2011.

Nora, Pierre. "Between Memory and History: Les Lieux de Mémoire." *Representations*, no. 26, spring 1989, pp. 7–24.

Norris, Margot. *Writing War in the Twentieth Century*. University of Virginia Press, 2000.

Parry, Jonathan. *The Politics of Patriotism: English Liberalism, National Identity and Europe, 1830–1886*. Cambridge University Press, 2006.

Payne, Matthew Thomas. *Playing War: Military Video Games after 9/11*. New York University Press, 2016.

Pederson, Joshua. "Speak, Trauma: Toward a Revised Understanding of Literary Trauma Theory." *Narrative*, vol. 22, no. 3, October 2014, pp. 333–353.

Pellerin, Cheryl. "Robots Could Save Soldiers' Lives, Army General Says." *Department of Defense News*, 17 August 2011. https://archive.defense.gov/news/newsarticle.aspx?id=65064.

Pennington, W. H. *Left of the Six Hundred!* Waterlow & Sons, 1887.

———. *Sea, Camp, and Stage: Incidents in the Life of a Survivor of the Light Brigade*. J. Arrowsmith, 1906.

Peterson, Jon. *Playing at the World: A History of Simulating Wars, People and Fantastic Adventures from Chess to Role-Playing Games*. Unreason Press, 2012.

Pick, Daniel. *War Machine: The Rationalization of Slaughter in the Modern Age*. Yale University Press, 1993.

Pitman, Roger, Scott P. Orr, Dennis F. Forgue, Jacob B. de Jong, and Jame M. Claiborn. "Psychophysiologic Assessment of Posttraumatic Stress Disorder Imagery in Vietnam Combat Veterans." *Archives of General Psychiatry*, vol. 44, no. 11, November 1987, pp. 970–975.

Ponting, Clive. *The Crimean War: The Story behind the Myth*. Chatto and Windus, 2004.

Poon, Angelia. *Enacting Englishness in the Victorian Period: Colonialism and the Politics of Performance*. Ashgate Publishing, 2008.

Porter, Patrick. *The Global Village Myth: Distance, War, and the Limits of Power*. Georgetown University Press, 2015.

Pötzsch, Holger. "Borders, Barriers and Grievable Lives: The Discursive Production of Self and Other in Film and Other Audio-Visual Media." *Nordicom Review*, vol. 32, no. 2, 2011, pp. 75–94.

———. "Selective Realism: Filtering Experiences of War and Violence in First- and Third-Person Shooters." *Games and Culture*, vol. 12, no. 2, 2017, pp. 156–178.

Prescott, Anna T., James D. Sargent, and Jay G. Hull. "Meta-Analysis of the Relationship between Violent Video Game Play and Physical Aggression over Time." *Proceedings of the National Academy of Sciences*, vol. 115, no. 40, 2018, pp. 9882–9888.

Prince, Stephen. "Celluloid Heroes and Smart Bombs: Hollywood at War in the Middle East." In *The Media and the Persian Gulf War*, edited by Robert E. Denton Jr. Praeger, 1993, pp. 235–256.

Prior, Denis. "Distant Thunder." In *Operation Homecoming: Iraq, Afghanistan, and the Home Front in the Words of U.S. Troops and Their Families*, edited by Andrew Carroll. Random House, 2006, pp. 25–37.

Prost, Antoine. "Monuments to the Dead." In *Realms of Memory: The Construction of the French Past*, vol. 2: *Traditions*, edited by Pierre Nora. Columbia University Press, 1997, pp. 307–332.

Puc, Samantha. "Space Invaders: Tomohiro Nishikado Reveals Why the Main Villains Were Octopuses." *CBR.com*, 20 August 2020. https://www.cbr.com/space-invaders-octopus-villains-war-worlds.

Reid, Julian. *The Biopolitics of the War on Terror: Life Struggles, Liberal Modernity and the Defence of Logistical Societies*. Manchester University Press, 2006.

———. "The Vulnerable Subject of Liberal War." *South Atlantic Quarterly*, vol. 110, no. 3, summer 2011, pp. 770–779.

Riegler, Thomas. "On the Virtual Frontlines: Video Games and the War on Terror." In *Videogame Cultures and the Future of Interactive Entertainment*. Inter-Disciplinary Press, 2010, pp. 53–62.

Ristelhueber, Sophie. *Eleven Blowups*. Bookstorming, 2006.

———. *Every One*. Arlogos Gallery, 1994.

———. *Fait: Koweit 1991*. Errata Editions, 2008. (Originally published 1992.)

———. *Operations*. Thames & Hudson, 2009.

Ritvo, Harriet. *The Animal Estate: The English and Other Creatures in the Victorian Age*. Harvard University Press, 1987.

Robbins, Alan Pitt. *Gilbert and Sullivan Operas*. Avon Publishing, 1950.

Robbins, Richard E., director. *Operation Homecoming: Writing the Wartime Experience*. Documentary Group, 2007.

Rossiter, Alicia Gill, and Sharlene Smith. "The Invisible Wounds of War: Caring for Women Veterans Who Have Experienced Military Sexual Trauma." *Journal of the American Association of Nurse Practitioners*, vol. 26, no. 7, July 2014, pp. 364–369.

Rothberg, Michael. "Preface: Beyond Tancred and Clorinda—Trauma Studies for Implicated Subjects." In *The Future of Trauma Theory: Contemporary Literary Theory and Cultural Criticism*, edited by Gert Buelens, Sam Durrant, and Robert Eaglestone. Routledge, 2014, pp. xi–xvii.

Ruskin, John. "Lectures on Art Delivered before the University of Oxford in Hilary Term, 1870." In *The Complete Works of John Ruskin*, vol. 20, edited by E. T. Cook and Alexander Wedderburn. George Allen and Unwin, 1903–1912, pp. 13–166.

———. "Sesame and Lilies." In *The Complete Works of John Ruskin*, vol. 19. Edited by E. T. Cook and Alexander Wedderburn. George Allen and Unwin, 1903–1912, pp. 5–192.

———. *Unto This Last: Four Essays on the First Principles of Political Economy*. John Wiley & Sons, 1885.

Russell, William Howard. *The Crimean War as Seen by Those Who Reported It*. Edited by Angela Michelli Fleming and John Maxwell Hamilton. Louisiana State University Press, 2009.

Russert, Tim. "Interview with Vice President Dick Cheney." *Meet the Press*, 14 September 2003. http://www.nbcnews.com/id/3080244/ns/meet_the_press/t/transcript-sept.

Scarry, Elaine. *The Body in Pain: The Making and Unmaking of the World*. Oxford University Press, 1985.

Semino, Elena. "Linguist Elena Semino Warns of the Negative Impact of War Metaphors on Cancer Patients." Universitat Internacional de Catalunya, 3 May 2019. http://www.uic.es/en/news/elena-semino-negative-impact-metaphors-cancer.

Shank, Nathan. "Productive Violence and Poststructural Play in the Dungeons and Dragons Narrative." *Journal of Popular Culture*, vol. 48, no. 1, 2015, pp. 184–197.

Simons, John, and John Louis Lucaites. *In/Visible War: The Culture of War in Twenty-First-Century America*. Rutgers University Press, 2017.

Simpson, William. "The Charge of the Light Brigade, Balaclava, 25th October, 1854." *Prints, Drawings and Watercolors from the Anne S. K. Brown Military Collection*. Brown Digital Repository. Brown University Library. (Originally published 1854.) https://repository.library.brown.edu/studio/item/bdr:232915/.

Sontag, Susan. *Regarding the Pain of Others*. Picador, 2003.

Speer, Isaac. "Reframing the Iraq War: Official Sources, Dramatic Events, and Changes in Media Framing." *Journal of Communication*, vol. 67, 2017, pp. 282–302.

Spilsbury, Julian. *The Thin Red Line: An Eyewitness Account of the Crimean War*. Weidenfeld & Nicolson, 2005.

Stahl, Roger. "Digital War and the Public Mind: *Call of Duty* Reloaded, Decoded." In *In/Visible War: The Culture of War in Twenty-First-Century America*, edited by John Simons and John Louis Lucaites. Rutgers University Press, 2017, pp. 143–158.

———. *Through the Crosshairs: War, Visual Culture, and the Weaponized Gaze*. Rutgers University Press, 2018.

———. "What the Drone Saw: The Cultural Optics of the Unmanned War." *Australian Journal of International Affairs*, vol. 67, no. 5, 2013, pp. 659–674. http://dx.doi.org/10.1080/10357718.2013.817526.

Standage, Tom. *The Victorian Internet: The Remarkable Story of the Telegraph and the Nineteenth Century's On-Line Pioneers*. Berkley Books, 1999.

Stargardt, Nicholas. *The German Idea of Militarism: Radical and Socialist Critics, 1866–1914*. Cambridge University Press, 1994.

Stauble, Katherine. "Of Fact and Fiction: Sophie Ristelhueber's *Fait*." National Gallery of Canada, 26 November 2015. https://www.gallery.ca/magazine/exhibitions/of-fact-and-fiction-sophie-ristelhuebers-fait.

Stubblefield, Thomas. *Drone Art: The Everywhere War as Medium*. University of California Press, 2020.

———. "In Pursuit of Other Networks: Drone Art and Accelerationist Aesthetics." In *Life in the Age of Drone Warfare*, edited by Lisa Parks and Caren Kaplan. Duke University Press, 2017, pp. 195–219.

Summers, Anne. *Angels and Citizens: British Women as Military Nurses, 1854–1914*. Threshold Press, 2000.

Swofford, Anthony. *Jarhead: A Marine's Chronicle of the Gulf War and Other Battles*. Scribner, 2003.

Sylvester, Christine. *Masquerades of War*. Routledge, 2015.

Tate, Trudi. "On Not Knowing Why: Memorializing the Light Brigade." In *Literature, Science, Psychoanalysis, 1830–1970*, edited by Helen Small and Trudi Tate. Oxford University Press, 2003, pp. 160–180.

Taylor, John. *Body Horror: Photojournalism, Catastrophe and War*. New York University Press. 1998.

Tennyson, Alfred, Lord. "The Charge of the Heavy Brigade." In *The Complete Poetical Works of Alfred, Lord Tennyson, Poet Laureate*. H. B. Nims and Company, 1885, p. 373.

———. "The Charge of the Light Brigade." In *The Complete Poetical Works of Alfred, Lord Tennyson, Poet Laureate*. H. B. Nims and Company, 1885, pp. 147–148.

———. "Locksley Hall." In *The Complete Poetical Works of Alfred, Lord Tennyson, Poet Laureate*. H. B. Nims and Company, 1885, pp. 59–62.

———. "Maud." In *The Complete Poetical Works of Alfred, Lord Tennyson, Poet Laureate*. H. B. Nims and Company, 1885, pp. 129–142.

Tolstoy, Leo *The Sebastopol Sketches*. Edited by David McDuff. Penguin, 1986.

Tosh, John. *Manliness and Masculinities in Nineteenth-Century Britain*. Pearson Longman, 2005.

Tucker, Herbert F. *Tennyson and the Doom of Romanticism*. Harvard University Press, 2014.

Turner, Brian. *Here, Bullet*. Alice James Books, 2005.
Turner, Jon. "Untitled." In *Warrior Writers: Re-Making Sense*, edited by Lovella Calica. Iraq Veterans against the War, 2008, pp. 119–120.
United Nations International Children's Emergency Fund (UNICEF). "Patterns in Conflict: Civilians Are Now the Target." n.d. https://www.unicef.org/graca/patterns.htm.
"U.S. Military Spending/Defense Budget 1960–2020." *Macrotrends*, n.d. https://www.macrotrends.net/countries/USA/united-states/military-spending-defense-budget.
Usherwood, Paul, and Jenny Spencer-Smith. *Lady Butler, Battle Artist, 1846–1933*. Alan Sutton Publishing, 1987.
Vågnes, Øyvind. "Drone Vision: Towards a Critique of the Rhetoric of Precision." *Krisis: Journal for Contemporary Philosophy*, no. 1, 2017, pp. 8–17. https://krisis.eu/drone-vision.
Vanden Bossche, Chris R. "Realism versus Romance: The War of Cultural Codes in Tennyson's 'Maud.'" *Victorian Poetry*, vol. 24, no. 1, spring, 1986, pp. 69–82.
Vandermeulen, Bruno, and Danny Veys, editors. *Imaging History: Photography after the Fact*. ASA Publishers, 2011.
Varhola, Michael O. "Gary Gygax's Foreword to *Little Wars*." Independent Tabletop Game Network, n.d. https://d-infinity.net/skirmisher/gary-gygaxs-foreword-hg-wells-little-wars.
Virilio, Paul. *Speed and Politics: An Essay on Dromology*. Semiotext(e), 1977.
———. *War and Cinema: The Logistics of Perception*. Translated by Patrick Camiller. Verso, 1984.
Waddington, Patrick. *"Theirs but to Do and Die": The Poetry of the Charge of the Light Brigade at Balaklava, 25 October 1854*. Astra Press, 1995.
Walberg, Eric. *Postmodern Imperialism: Geopolitics and Great Games*. Clarity Press, 2011.
Weaver, Mike. "Roger Fenton: Landscape and Still Life." In *British Photography in the Nineteenth Century: The Fine Art Tradition*, edited by Mike Weaver. Cambridge University Press, 1989, pp. 103–120.
Weber, Julia. "Game Changer? On the Epistemology, Ontology, and Politics of Drones." *Behemoth*, vol. 8, no. 2, 2015, pp. 1–11.
Welland, Julia. "Violence and the Contemporary Soldiering Body." *Security Dialogue*, vol. 48, no. 6, 2017, pp. 524–540.
Wells, H. G. *An Experiment in Autobiography: Discoveries and Conclusions of a Very Ordinary Brain*. Macmillan, 1934.
———. *Floor Games*. Frank Palmer, 1911.
———. *Little Wars*. Edited by Nigel Lepianka and Deanna Stover. *Scholarly Editing*, vol. 38, 2017. (Originally published 1913.) http://scholarlyediting.org/2017/editions/littlewars/intro.html#inlinenote31.
———. *The New Machiavelli*. Bodley Head, 1911.
———. *Seven Famous Novels by H. G. Wells*. Alfred A. Knopf, 1934.
———. *The War of the Worlds*. Charles Scribner's Sons, 1924.
Whitehead, Neil L., and Sverker Finnström, editors. *Virtual War and Magical Death*. Duke University Press, 2013.
Wickens, Glen. "Hardy, Militarism and War." In *Thomas Hardy in Context*, edited by Phillip Mallett. Cambridge University Press, 2013, pp. 415–424.
Williams, Alison J. "Disrupting Air Power: Performativity and the Unsettling of Geopolitical Frames through Artworks." *Political Geography*, no. 42, 2014, pp. 12–22.

Williams, Kayla. *Love My Rifle More Than You: Young and Female in the U.S. Army.* W. W. Norton, 2005.

Winthrop-Young, Geoffrey. "Drill and Distraction in the Yellow Submarine: On the Dominance of War in Friedrich Kittler's Media Theory." *Critical Inquiry*, vol. 28, no. 4, summer 2002, pp. 825–854.

Yarrington, Alison. *The Commemoration of the Hero 1800–1864: Monuments to the British Victors of the Napoleonic Wars.* Garland, 1988.

Young, Allan. *The Harmony of Illusions: Inventing Post-Traumatic Stress Disorder.* Princeton University Press, 1995.

Younge, Gary, and Jon Henley. "Wimps, Weasels and Monkeys—the US Media View of 'Perfidious France.'" *The Guardian*, 11 February 2003. https://www.theguardian.com/world/2003/feb/11/pressandpublishing.usa.

Zelizer, Barbie. "When War Is Reduced to a Photograph." In *Reporting War: Journalism in Wartime*, edited by Stuart Allan and Barbie Zelizer. Routledge, 2004, pp. 115–135.

Index

Note: Page numbers in italics refer to figures.

About the Author

MARTIN A. DANAHAY is a professor of English at Brock University, Canada. He has published widely on Victorian literature and culture, including such topics as gender and work, the working-class body, the Arts and Crafts movement, and H. G. Wells and eugenics. He is the author of *Gender at Work in Victorian Culture: Literature, Art and Masculinity* and *A Community of One: Masculine Autobiography and Autonomy in Nineteenth-Century Britain.*